Advanced Smart Grids for Distribution System Operators

Advanced Smart Grids Set

coordinated by
Nouredine Hadjsaïd and Jean-Claude Sabonnadière

Volume 1

Advanced Smart Grids for Distribution System Operators

Marc Boillot

iSTE

WILEY

First published 2014 in Great Britain and the United States by ISTE Ltd and John Wiley & Sons, Inc.

ISTE Ltd
27-37 St George's Road
London SW19 4EU
UK

www.iste.co.uk

John Wiley & Sons, Inc.
111 River Street
Hoboken, NJ 07030
USA

www.wiley.com

Library of Congress Control Number: 2014953030

British Library Cataloguing-in-Publication Data
A CIP record for this book is available from the British Library
ISBN 978-1-84821-737-9

Contents

Foreword

In most countries, the concept of smart grids is becoming increasingly significant, mostly driven by societal concerns such as reliability, cyber and physical security of supply, transmission and delivery of energy, as well as climate change and aging assets. These concerns are expressed in terms of objectives such as those set by the European Union (EU) through the "climate and energy package" adopted in 2009 for 2020, which consists of reducing CO_2 emissions by 20% compared to 1990, increasing the share of renewable energy to 20% and increasing energy efficiency by 20%. The two first objectives are binding targets adopted by all EU member states. Making the demand more responsive to the condition of the power system is also needed in order to accommodate the anticipated changes brought about by larger deployment levels of renewable generation technologies. Worldwide, other countries have set their own objectives depending on their needs and priorities. As a result, through strong regulation incentives, a remarkable development of renewable energy sources (RES) has been observed globally particularly in wind and solar energy. Plug-in hybrid electric vehicles (PHEV) are also on the rise in the global car industry.

The vast majority of these sources are connected to the electrical grid at either transmission or distribution levels.

Electrical networks are undergoing tremendous changes in order to accommodate this evolution that is in most cases very dynamic. However, for some countries, such as France, 95% of these sources are located at the distribution level, traditionally operated as a radial mode (unidirectional power flows) as little or no energy sources existed there in the past. In fact, unlike transmission grids which are already "smart" (seen as the backbone of the entire electrical system with embedded monitoring, control and protection technologies), distribution networks have thus far received much less attention in terms of smart technologies. However, with the ongoing aforementioned changes, distribution networks are in the front line with the development of RES, PHEV as well as end-users, who are expected to play a more active role in this new energy paradigm. They are becoming *prosumers* (producers/ consumers).

Facing these changes requires the development and integration of enabling technologies and energy services that are based on new energy technologies while taking advantage of more information and communication technologies. The entire energy chain is at stake here: smart meters, demand response, storage, smart substations, self-healing, advanced observability and control functions, advanced communication and big data processing capabilities across the network, and the portfolio of value-added functions that it may bring about, and so on.

Distribution companies and distribution system operators (DSOs) in particular are facing unprecedented challenges in their networks. In addition, they have to respond to them in an increasing number of ways, prompting concerns of the quality of supply among network users, fast development of new uses for energy supplies and effective management of aging electric utility assets, occurring very often in an unstable regulation landscape.

This book is precisely targeting the changes that are rapidly occurring at the distribution level and the role of DSOs in the development of the smart grid concept. It gives a remarkable insight into the industry perspective on several aspects such as necessary technology, operational and planning stakes, examples of value chain of some smart grid pilot projects worldwide with original view points on learned lessons and key findings of smart grids. This book undoubtedly contains very useful knowledge on smart grid evolution in the realm of distribution networks – a great resource for all readers interested in this exciting subject.

We hope this book will receive a warm welcome from the community of researchers and engineers from industry as well as academia, all of whom are contributing in small and not-so-small ways to the (r)evolution of the smart distribution networks of the future.

Miroslav BEGOVIC
President
Institute of Electrical and Electronics Engineers (IEEE)
Power and Energy Society (PES)
October 2014

Preface

Smart grids are systems that are obtained by superposing information and telecommunication networks on electric power networks.

Their purpose is to integrate intermittent renewable energy sources (RES) (such as solar photovoltaics and wind) and new applications for electric power (such as electric vehicles) in the best safety conditions, while minimizing investments into reinforcing traditional power networks.

For this reason, distribution system operators (DSOs) develop intelligent networks by integrating various technologies, such as sensors, smart meters, reinforced chains of information transmission and exchange, real-time analysis, decision-support softwares, automation and remote-controlled functions, etc.

For 15 years, DSOs made important investments in medium-voltage networks, which led to improving the service quality and greatly lowering the average outage time for customers. These investments also made possible the growth of the share of renewable energy sources, in particular of those known as intermittent.

The challenge over the next few years is to modernize low-voltage networks, as has been previously done with medium-voltage networks.

Marc BOILLOT
October 2014

Acknowledgments

The author would like to acknowledge all the contributors who made possible the accomplishment of this project.

Nouredine Hadjsaid and Jean-Claude Sabonnadière for their stimulation and support; without them, this book would not have seen the light of day.

Alain Doulet for his knowledge of the history of distribution networks, his competences on the smart grids and his ability to anticipate the future.

All colleagues from ERDF, in particular those from the different teams Smart Grids, Linky, Strategy and International projects, from the technical division and the IT division and finally the Regions which are involved with smart grids projects.

All people who, in Europe, in the United States and in Asia, contributed to provide a worldwide scale to smart grids projects.

All colleagues and friends from the G3-PLC Alliance who worked with success toward the standardization and the promotion of the G3-PLC to the DSOs and all potential users worldwide.

List of Figures

List of Acronyms

6LowPAN	Network Layer Protocol of the OSI model
ACER	Agency for the Cooperation of Energy Regulators
AD	active demand
ADEME	*Agence de l'Environnement et de la Maîtrise de l'Energie* (French agency for the environment and control of energy)
ADSL	asymmetric digital subscriber line
ADVANCED	Active Demand Value and Consumers Experiences Discovery
AENS	average energy not supplied
AMI	advanced metering infrastructure
AMM	automated meter management
AMR	automated meter reading
ARIB	frequency band (155–403 kHz) for PLC communication in Japan
ARRA	American Recovery and Reinvestment Act
ASK	amplitude-shift keying
ASUI	average service unavailability index
ATEE	*Association Technique Energie Environnement* (French technical association for energy and the environment)
BAU	business as usual
BEMS	Building Energy Management System
CAES	compressed air energy storage

CAPEX	capital expenditure
CEATEC	Combined Exhibition of Advanced Technologies trade show in Japan
CEM	Clean Energy Ministerial
CEMS	Community Energy Management System
CENELEC-A	frequency band A (35–91 kHz) for PLC communication in Europe
CEO	Chief Executive Officer
CIGRE	*Conférence Internationale des Grands Réseaux d'Electricité* (Internatoinal Conference on Large Electricity Networks)
CHP	combined heat and power
CO_2	carbon dioxide
CSI	commercially sensitive information
DBPSK	differential binary PSK
DCs	data concentrators
DCPR	distribution price control review
DCPS	digital controlled primary substations
DER	distributed energy ressources
DG	distributed power generation
DGCIS	*Direction Générale de la Compétitivité, de l'Industrie et des Services;* this Direction has been transformed in September 2014, into *DGE Direction Générale des Entreprises* (French business executive)
DMS	distribution management system
DOE/EIA	Department of Energy/Energy Information Administration
DSM	demand-side mnagement
DSO	distribution system operators
DQPSK	differential quadrature DPSK
EC	European Commission
EET	extreme energy transition
EJP	*Effacement "Jours de Pointe"* (load management)
EDF	*Electricité de France* (French electricity company)
EDSO	European Distribution System Operators

ENTSO-E	European Network of Transmission System Operator – Electricity
ENTSO-G	European Network of Transmission System Operator – Gas
ENWL	Electricity North West Limited
EPRI	Electricy Power Research Institute
ERDF	*Electricité Réseau Distribution France* (French electricity distribution network)
EU	European Union
EU FP7	EU's Seventh Framework Programme for Research
EV	electric vehicle
EWE	Energieversorgung Weser-Ems AG
FCC	frequency band (150–487.5 kHz) for PLC communication in the USA and other countries
FEMS	Factory Energy Management System
FSK	frequency-shift keying
GHG	greenhouse gas
GIS	geographical information system
GPRS	General Packet Radio Service
GSM	Global System for mobile Communication
GW	Giga Watt
HEMS	Home Energy Management System
HV	high voltage
ICT	information and communication technologies
IEA	International Energy Agency
IEOD	information exchange and operating devices
IEC	International Electrotechnical Commission
IEEE	Institute of Electrical and Electronics Engineers
IFFT	inverse fast Fourier transformation
IS	information systems
ISGAN	International Smart Grid Action Network
ITU	International Telecommunication Union
JRC	Joint Research Center
KEPCO	Korea Electric Power Corporation

KPI	key performance indicator
KSGI	Korea Smart Grid Institute
LAN	local area network
LRE	Linky radio emitter
LQS	low quality of supply-customers
LV	low voltage
MAC	media access control layer of the OSI model
MEMS	MicroElectroMagnetic Systems
METI	Ministry of Economy, Trade and Industry
MV	medium voltage
NEDO	New Energy and Industrial Technology Development Organization
NOC	Network Operation Center
NPV	net present value
OECD	Organisation for Economic Co-operation and Development
OH	off-peak hours
OFDM	orthogonal frequency division multiplexing
O&M	operation and maintenance
OPEX	operational expenditure
PDN	public distribution network
PH	peak hours
PHEV	plug-in hybrid electric vehicle
PHY	physical layer of the OSI model
PLC	power line carrier
PSK	phase-shift keying
PV	photovoltaic
R&D	research and development
RCD	remote control device
REDOX	reduction and oxidation reactions electro-chemical batteries
REMS	retail energy management system
RES	renewable energy sources
RF	radio frequency

ROUTE B	route for communications downstream the meter
RSP	renewable portfolio standards
RTU	remote terminal unit
RTE	*Réseau de Transport d'Electricité* (Electricity transport network)
RWE	Rheinisch-Westfälisches Elektrizitätswerk AG
SAIDI	system average interruption duration index
SAIFI	system average interruption frequency index
SCADA	supervisory control and data acquisition
SCE	Southern California Edison
SFSK	spread frequency shift keying
SG	steady growth
SGCC	State Grid Corporation of China
SME	small and medium enterprises
SMIs	small and medium industries
SNMP	Simple Network Management Protocol
SNR	signal-to-noise ratio
STN	switched telephone network
TIC	tele-information client
TFTP	Trivial File Transfer Protocol
TSO	transmission system operators
USP	unique software package
VPP	virtual power plant
WAN	wide area network

Welcome to "Advanced Smart Grids"

This book on advanced smart grids is divided into eight chapters.

Chapter 1: Distribution System Operators in a Changing Environment. This introductory chapter presents the process of the energy transition that is under way in many regions of the world to face the increase in demand and accompany the development of renewable energy sources (RES). The distribution system operators (DSOs) play a key role in the electric system. They develop intelligence at the heart of the distribution network and act as market facilitators. They make use of existing and new energy technologies, as well as information and telecommunication technologies that support these energy technologies.

Chapter 2: The Existing Distribution Networks: Design and Operation. We emphasize the principles that guide the development of electricity distribution networks. Various technical approaches were implemented worldwide for the amount of choice and the value of voltage levels, as well as for the medium-voltage (MV) neutral point treatment and for the required level of quality. France, for example, reviewed a lot of its technological choices between 1960 and 2010: changing 15/20 kV voltage, changing neutral point treatment, shifting toward underground (MV) and low-voltage (LV) networks, then orientating its actions toward improving quality and desensitizing climate hazards.

Chapter 3: Main Divers and Functions of Advanced Smart Grids. This chapter presents the smart grids. The massive input of RES promotes the development of network observability, in real time, and reinforces its control. The goal is to optimize the costs, while allowing the network to increase its RES carrying capacity. To reach this objective, it is appropriate to take advantage of solutions for dynamic management of constraints. The secondary substation is an essential element as it has the potential to become a privileged point of observability, as well as communication node between information technologies (IT) and downstream uses. Managing the network of tomorrow will involve a better understanding of the state of the network in real time and with forecasts. Primarily, smart grids are used for the operation and development of the network, the dynamic management of constraints and distinctions between flexibility levers.

Chapter 4: Metering: a Core Activity of the Distribution System Operators. In this chapter, our main focus is on the smart meter: advanced metering infrastructure (AMI). DSOs are in an optimal position to deploy and manage the metering infrastructure that forms part of the network. Smart metering systems have become a standard that provides solutions to changes in regulation, improves customer satisfaction, makes the energy transition possible and improves distribution performance. Power line carrier (PLC) technology is presented in its most advanced version: the G3-PLC. The data from the meters, supplemented with network events, are capable of detecting cases of low quality supply to customers, following supply quality in any given geographical area, monitoring power quality, etc. Smart meters thereby contribute to the development of smart grids.

Chapter 5: Focus on Flexibility Options. This chapter focuses on the flexibility options and how demand is managed. DSOs act as market facilitators. They will be able to buy "flexibility" solutions from market players, alternatively or complementary to network reinforcement. Among the options, we find, notably, management of the location of RES, local peak management, active management of generation, reactive power management, etc. For illustrative purposes, we present the smart meter as a facilitator of flexibility: with this new tool, energy suppliers will be able to provide innovative pricing offers to limit the local peak power and optimize energy consumption. The smart meter, as a bridge between the network and the customer, makes data available to the market players (suppliers, aggregators, customers, etc.) in order to allow them to adapt their activity.

Chapter 6: Pilot Projects and Use Cases. In this chapter, we present some of the numerous smart grid demonstration projects conducted around the world to address major technological themes. The *use cases* methodology was created to equip these smart grids projects (description of business processes, IT functions, feedback of experience, etc.). The case of the European project Grid4EU is presented with six demonstrators, as well as four other cases from the ISGAN Case Book on Demand-Side Management.

Chapter 7: Smart Grids Are the Future for DSOs. This chapter aims to identify the conditions that will allow DSOs to develop smart grids. Smart grids will require new capacities: *big data*, forecasting of local generation and demand, management of telecom and IT infrastructures, and shared interfaces with the operators of electric systems, among others. The development of smart grids provides a unique opportunity for DSOs: a high-tech image alongside technological innovations, DSOs as key players in the evolution of the network, and responsibility for the societal and environmental expectations of customers and market players.

Chapter 8: Key Findings. We gather here the primary conclusions of this book: smart grids are first and foremost the current and future power networks, superposed on a communication network and a processing and monitoring system. The role of the DSOs becomes central in the distribution of responsibilities at the core of the electric system: the DSO ensures the stability of the voltage level at the local scale. Forecast management becomes a genuine job for wind and solar generation, which leads to anticipated constraints. The DSO implements flexibility in order to remove these constraints. The DSO is not a load-shedding player: it makes possible the emergence of new flexibility devices. The generalized deployment of smart meters provides several advantages for market players and for customers. If they wish, the customer can become a player in their own right, and influence their own energy consumptions. Smart metering also aims to allow the DSO to monitor the LV network and control it better. Smart grids represent a real industrial opportunity and reinforce spectacularly the attractiveness of the DSOs.

1

Distribution System Operators in a Changing Environment

1.1. Energy policies promoting the energy transition

During the last three decades, strong economic growth and expanding populations have lead to a significant increase in global energy demand. For the next three decades, many forecasts unanimously predict that this increase will continue at this pace. Also, because of the economic growth of China and India, the rate is accelerated in non-OECD (organization for economic co-operation and development) economies.

To support the energy demand, global net electricity generation has increased quickly from 1990 to 2010 and will supply an increasing share of the total demand from 2010 to 2040 as shown in Figure 1.1.

Electricity consumption by end-users is expected to grow faster than the use of other energy sources due to the increase in the standard of living and a higher demand for home appliances and electronic devices. This is also true with the expansion of professional sector's needs such as hospitals, office buildings, commercial services, shopping malls, etc.

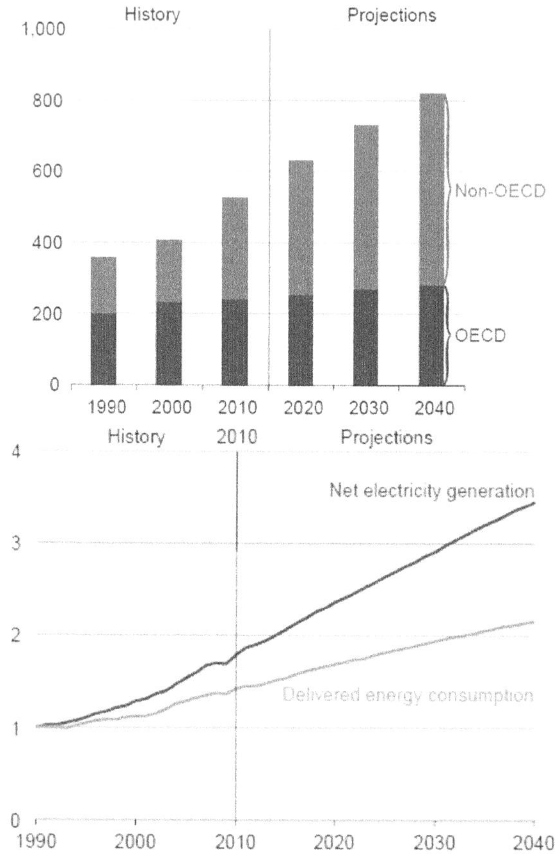

Figure 1.1. *World total energy consumption 1990–2040 (quadrillion btu)*[1] *and world electricity generation (index, 1990 = 1)*[2]*. For a color version of the figure, see www.iste.co.uk / boillot / smartgrids.zip*

Combinations of primary energy sources to produce electricity will be evolving in a significant way over the next three decades:

1 Source EIA – International Energy Outlook 2013.
2 Source EIA – International Energy Outlook 2013.

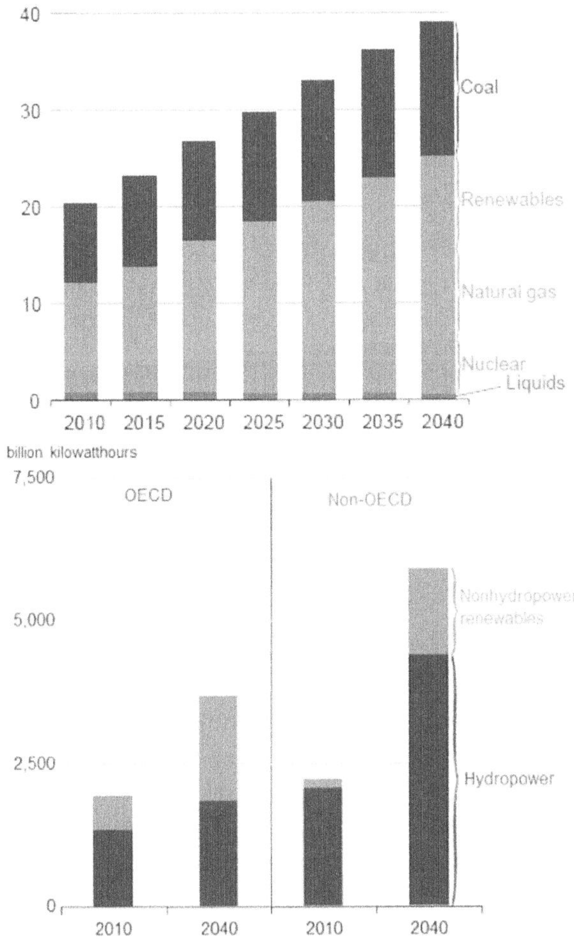

Figure 1.2. *World electricity generation by fuel 2010–2040 (trillion kWh) and world electricity generation from renewable energy sources 2010 and 2040*[3]. *For a color version of the figure, see www.iste.co.uk/boillot/smartgrids.zip*

In particular, according to US Department of Energy/Energy Information Administration (DOE/EIA)

3 Source: US Energy Information Administration (DEO/EIA) – International Energy Outlook 2013.

Reference Case projections, the renewable share of these combinations will increase from 21 to 25% – the world fastest growing source of electric power. Worldwide hydropower will account for 52% of the total increment and wind generation for 28%, with large differences between regions and countries:

– most renewable energy in OECD countries is expected to come from non-hydroelectric energy, because all resources have already been developed (except Canada and Turkey);

– in non-OECD countries, hydroelectric power is expected to be a dominant source of growth (in particular Brazil, China and India). Nevertheless, growth rates for wind power electricity will also be high. Particularly in China, where wind generated electricity should go from 6% in 2010 to 26% in 2040 (45–637 TWh of expected generated energy respectively).

Facing the challenge of a growing demand of energy, many regions of the world are engaged in a dymanic phase of energy transition. The production of electricity from renewable sources and, particularly, intermittent sources, is increasing in many regions. By 2012, more than 280 GW of wind farms and 100 GW solar photovoltaic (PV) are installed worldwide. The International Energy Agency (IEA) forecasts on a shorter term basis that the evolution will continue with the installation of +230 GW of wind power and +210 GW of solar PV by 2017.

Many governmental organizations encourage the development of sustainable transportation facilities (train, buses, tramway, etc.), and car manufacturers are now offering a wide range of plug-in hybrids and other electric vehicles (in December 2012, around 180,000 plug-in electric vehicles (EVs) were already on the road[4]).

4 Source IEA – Global EV Outlook 2013.

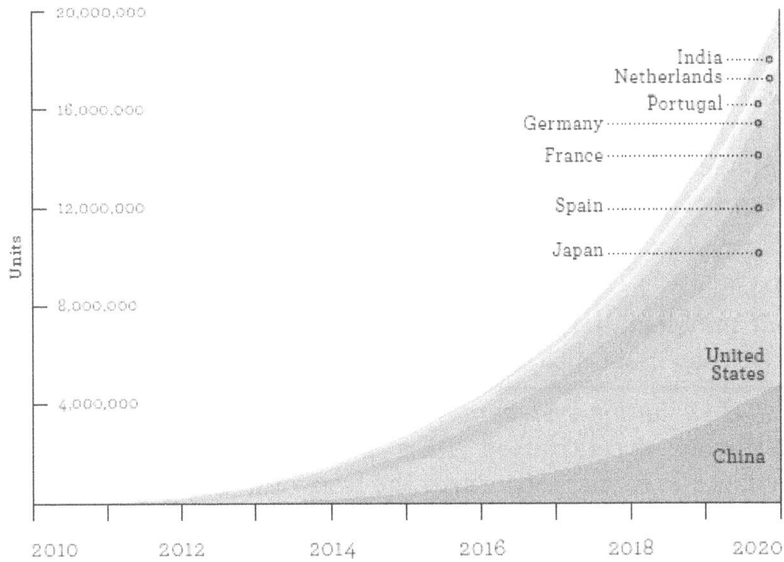

Figure 1.3. *Project of the evolution of EV throughout the world (plug-in and hybrid plug-in). Source: IEA – Global EV Outlook 2013. For a color version of the figure, see www.iste.co.uk/boillot/smartgrids.zip*

Last but not least, consumers are changing their attitude toward energy savings. The massive roll-out of electric smart meters will permit the development of energy conservation services. More than 80 million smart meters were already deployed worldwide by December 2013 including 46 million in the USA[5]. This number is expected to reach 100 million meters by the end of 2014 according to IHS Inc[6], and 1 billion meters by the end of 2020 according to Pike Research[7].

5 http://www.edisonfoundation.net.

6 http://www.cepro.com/article/100_million_smart_meters_to_be_installed_worldwide_in_2014.

7 http://www.navigantresearch.com/newsroom/the-installed-base-of-smart-meters-will-surpass-1-billion-by-2022.

The changes in generation means and consumption trends will impact energy systems worldwide:

– *Producers* will have to alter their business models in order to make their investments in existing generation facilities profitable, as well as to optimize operational management of energy combinations that increasingly integrate intermittent renewable energy sources (RES);

– *Transmission system operators (TSOs)* will have to anticipate the risks of an unbalanced supply-demand ratio that may lead to a decrease in frequency and potential black-outs; they must also develop interconnections;

– *Distribution system operators (DSOs)* will have to connect massively decentralized RES generation, electric vehicle recharge stations, modernize the networks and deploy smart grid technologies including metering systems;

– *Energy suppliers* will have to reevaluate their offers and services in response to consumers' expectations in the context of an increasingly competitive environment (progressive market opening, with the end of regulated tariffs).

The energy transition makes a major impact for DSOs, insofar as intermittent RES generation installations are predominantly connected to distribution networks. For instance, in France, at the end of 2013, 94% of RESs installations, around 300,000, were connected to the distribution network and represented a total of 11.4 GW[8].

To keep up with current energy volume, the total capacity of RES installations must be nearly five times greater than the capacity of current centralized thermal or nuclear generation sites. Indeed, the average running times for wind and solar power stations are around 2,000 and 1,000 h per year, respectively (average in France), while baseline generation times for a thermal or nuclear station can reach

8 ERDF Source – key figures 2013.

7,000–8,000 h per year. It is important to remind that wind and solar PV generation is not guaranteed and that the correlation with demand is generally low, depending on geographical location and types of usage.

EU DSOs landscape

The electricity distribution business in Europe includes more than 2,400 companies, which serve around 260 million connected customers supplying move than 500 million people, operating 10 million km of power line, distributing around 3,000 TWh a year and directly employing more than 240,000 people[9].

In most European countries, intermittent energy generation is developing very fast, leading to a total installed capacity of 106 GW of wind and 70 GW of PV by the end of 2012[10]. The vast majority of these plants are connected to distribution grids. Together with the development of active demand and electric vehicles, *this will lead to a pivotal transformation of the role of the DSOs.*

A real challenge for electric systems – a paradigm shift for distribution networks. In yesterday's market, the distribution networks were often designed to be operated radially in order to distribute electricity from HV/MV substations connected to transmission level, down to the end-user consumers. With the energy transition, tomorrow's electricity distribution network operation and management will change. The distribution networks will have to manage more complex interlinked networks mixing generation and demand with much higher variations and reverse flows from distribution to transmission networks. Also, new market players are developing, such as load curtailers, virtual power plant operators and aggregators, etc.

---

9 Article IEEE P&E magazine – Future of Power Distribution, European perspectives.
10 Source: Observ'Er : Etat des energies renouvelables en Europe – 2013.

At the same time, the basic principles of electric systems have not changed. Electricity must be generated at the same time it is consumed, whereas only small amounts can be stored. Also, voltage and frequency levels must be kept within prescribed limits to ensure the security and stability of electricity supply.

Smart grids are not an option, they are a necessity. To tackle the challenges of energy transition, electricity networks will need to be more reactive and flexible to ensure the security and stability of the system, and also enhance interactions between market players. Smart grids will contribute to those objectives by combining advanced electricity network technologies with information and communication technologies.

The main principles of smart grids are:

1) *collect data* on networks due to sensors and remote controlled devices (e.g. smart meters);

2) *analyze* the status of the network on a short-time basis;

3) *maintain or improve* the quality of supply by developing advanced tools and strategies to handle various disturbances and faulty situation (from predictive maintenance to self-healing grid);

4) *anticipate* local generation from RES; simulate the constraints on the network (short-, medium- and long-term approaches) and find solutions to manage safely the flows of electricity;

5) *manage optimally* the interaction grid-plug-in hybrid electric vehicle (PHEV);

6) *enable the development* of energy conservation services: by giving information on electricity consumption and permitting to manage electricity uses;

7) *manage peak situations* and interact intelligently with the end-user (consum'actor).

The DSOs play a central role in the deployment of smart grids. In charge of reliable operations of the distribution grid, DSOs should act as enablers and facilitators of the market in order to:

– ensure uniformed and harmonized deployment of smart grids;

– enable the deployment of new services;

– contribute to the operation and control of new flexibilities (storage, peak shaving programs, management of capacities (production and demand), dynamic tariffs, etc.);

– provide data to the customers, suppliers and other market players and ensure its security and usability.

1.2. A new era of technological revolution

For over 120 years, the electric power distribution sector successfully resolved technical and financial challenges brought by the increasing demand accompanying economic and demographic growth.

Until recently, electric power distribution was essentially a capacity network, simple and robust, featuring a minimum of complex systems; its functions were limited to transferring energy from the upstream (high-voltage transmission network) to the downstream (customers). This robustness was partly due to the large amount of equipment whose reliability was easy to guarantee, often owing to its fundamental simplicity. It was also ensured by the need for operational safety, which inherently led to taking special precautions while introducing new information and telecomunication technologies (ICTs), because of the disruptive nature of the electromagnetic environment surrounding medium-voltage installations.

Nevertheless, new functions were added progressively: remote control for network breakers and switches, automation devices for limiting supply interruption to the customers during incidents and remote monitoring systems to lower operational costs. France, in particular, favored deploying digital controlled primary substations (DCPSs) since the beginning of 2000.

These developments transformed the simple electric power distribution grid into a system composed of three highly synergized levels: a power network, a communication network between key points of the power network and a centralized control and command system, and finally, an information processing and monitoring system. Every level is confronted with its own major difficulties:

– *the electric network does not undergo technical revolutions, however, its oldest equipment suffers from age.* Optimizing renewal is a major challenge. It is essential to operate installations as close as possible to their limit and to be aware of their lifetime expectancy. This shows the importance of real-time monitoring systems based on decentralized sensors;

– *the communication network is the key point to modernizing the electric power network.* Today, it may be based on various technologies: dedicated lines, power line carrier (that is using the electrical current as support for the signal), shortwave radio, microwave transmission, optical fiber, etc. This will allow connecting various sensors to a single central monitoring point;

– *the information and monitoring system must maintain its performance* despite the important volume of incoming data and find an equilibrium between centralized and decentralized intelligence.

To resolve these challenges, the DSOs will be able to take advantage of a series of new technologies that will need to be integrated with current or future technologies. McKinsey Global Institute recently published a study[11] identifying 12 technologies with the greatest breakthrough potential on the 2025 horizon. Some examples include energy-storage, Internet of Things and *big data*.

At the heart of DSOs core business, most notable breakthroughs will likely occur with microelectromagnetic systems (MEMS)-based sensors), nanotechnologies and power electronics.

Thus, advanced smart grids will utilize a combination of different technologies throughout integrated and interoperable solutions. Certain technologies are already accessible (smart meters and network automation), others may be here in less than 10 years (energy storage and smart household communicating devices).

11 May 2013 – Disruptive Technologies: Advances that will transform life business, and the global economy.

2

The Existing Distribution Networks: Design and Operation

The distribution network is an essential part of the electric power grid. These systems provide the connection between centralized generation facilities and vastly distributed consumer areas. If the high-voltage transmission grid transports large amounts of energy at high voltage to minimize losses, the distribution network serves areas on a smaller scale by branching out and progressively lowering voltage to values appropriate for consumers/appliances connected to the network.

Three major objectives guided (and are still guiding) the decision-making process behind developing and operating distribution networks:

1) *optimizing investment costs* in public infrastructure (balance between cost, performance and life-time expectancy);

2) *contributing to the reliability of the electric power grid* regardless of possible disruptions induced by weather conditions, human errors and actual variations of consumption and generation;

3) *ensuring the level of quality of distributed electricity* corresponds to the regulatory requirements (voltage plans and thermal limit ratings of various materials).

2.1. Above all, smart grids remain grids!

Before tackling the issue of intelligent power networks, it is important to explain what power networks are, why and how they were developed and how they are operated, to ultimately see that intelligence has already been progressively integrated into the networks by the distribution system operators (DSOs).

Various technical solutions have been created worldwide to develop power distribution infrastructures. First of all, this chapter aims to give some guidelines to understanding the principles behind the conception and operation of power distribution, before examining some cases in Europe, USA and Asia. Secondly, we will also look at the limits of the classic approach for developing distribution systems in the context of the new energy paradigm.

Deploying the smart grids' architecture relies on combining multiple levels of software and infrastructure in order to introduce communication, measurement, supervision and control into the network. Some of these functions and components existed for many years, and have already been integrated into modern networks (for example, in Europe) before the idea of smart grids.

The major categories of components and systems that can be found in smart grids are:

– *power infrastructure* (transmission and distribution): transmitting either alternative or direct current;

– *measurement tools*: gathering “raw” data about the status of the infrastructures;

– *communication infrastructure*: offering different levels of information exchange;

– *control and detection systems*: analyzing information from the measurement tools and detecting anomalies;

– *operating systems*: acting remotely on different system components.

2.2. The DSO, a player at the heart of the power system

The entire power system is based on three fundamental components: the generation, the network and the consumers.

As previously mentioned in the introduction, there are many sources of energy generation, notably fossil fuels (gas- and coal-fired power stations), nuclear and hydraulic (from run of the river to great dams). These generation sources, known historically as “centralized”, were developed in a relatively small number and were capable of supporting the needs of a large number of consumers. This approach reached an economic optimum by mutualizing costs. The generation sites are operated to cover every instance of power supply demand. Nevertheless, during the past 10 years, decentralized power generation (wind and solar PV) has also experienced a lot of development. This generation is characterized by low-voltage (LV), high intermittence, high spread throughout the territory and weak correlation with the electricity demand. The latter is particularly true for solar PV energy sources.

The consumers are downstream of the electricity supply network and are subdivided into different categories: electro-intensives (steel industries, railway transportation, etc.), enterprises and commerce, and residential consumers. Their needs are simple: uninterrupted access at competitive prices to the power and the electrical energy required by their activity. Some of the larger consumers have their own backup sources, such as fuel-powered generators. In light of the rise of renewable energy generation and sustainable development policies, businesses and residential consumers have also begun migrating toward small localized power generation, particularly in the form of solar panels. Generally, these generation facilities feed into the local power grid.

Electricity is difficult to store at an economically viable cost. Hence, electrical energy must be consumed as it is produced, or, conversely, produced as it is consumed. As a result, power supply networks have always been developed to provide an immediate liaison between producers and consumers. Above all, the managers of this liaison must maintain the balance between energy supply and energy consumption. They are at the heart of the electric power system.

Two major types of power systems currently exist: the transmission grid and the distribution network. The power transmission grid allows conducting of high-voltage electricity (400, 225, 90 kV in France) over great distances from the generation facilities to the geographical location of the consumers, with minimal losses by the Joule effect. The distribution network takes over for the latter part of this journey, at connection points known as primary substations, conducting electricity to the endpoint consumers by progressively lowering voltage to match the requirements of electric appliances. The distribution network is itself subdivided into two parts, the medium-voltage (MV) network (20 kV in France) and the LV network (400 V in France).

The transmission grid, despite covering a large geographic area, remains relatively small in terms of combined line length. It is the role of the distribution network to channel electricity to every consumer, regardless of their location within the covered territory. Hence, the overall line length of the distribution network is greatest of the two, and on the French territory, that length is 13 times greater than that of the transmission grid.

Traditionally, electricity systems were organized as a sort of monopoly, englobing everything from energy generation to its sale. Over the years, market liberalization processes were organized for energy generation and commercialization activities; nevertheless, network operation remained in a "regulated" domain, being part of a natural monopoly. In fact, there is little common interest in creating competition by multiplying electricity transmission and distribution infrastructures. On the contrary, it is beneficial to pool investments within a network in order to facilitate commercial activity for the players in the context of an unregulated generation and commercialization.

This period of market opening and unbundling of established players made clear the functions and goals of system operators, in particular of the DSOs:

– *maintain and modernize power distribution infrastructures* (power lines, transformers, etc.), whether they are privately or publicly owned;

– *develop and maintain a metering infrastructure*, keep track of energy consumed by each customer and generated by each producer;

– *guarantee the quality of power supply*;

– *guarantee the safety of the supply*;

– *provide indiscriminate access to the grid,* connect all new generation and consumption facilities, regardless of electricity supplier and customer;

– *facilitate implementation of competitive market mechanisms, while remaining a neutral player.*

Furthermore, the goal and missions of a DSO may vary depending on the country and size of the operator (from hundreds of thousands of customers to tens of millions).

Ultimately, under all circumstances, a DSO must deliver electricity (energy and capacity) to every customer within the conditions guaranteeing quality and safety. The DSO must use all means and technologies necessary to monitor the operation of the power grid under normal or disrupted operation, in order to make sure that no hazard could lead to an interruption of the energy supply service.

The recent changes in the conditions of operation of power grids, brought by the massive proliferation of renewable energy source (RES), new uses for electric power and new services, as well as an increased frequency of exceptional weather events (periods of great heat or cold, floods and storms), give even more importance to the role of the DSO in the world of electricity distribution, to the benefit of consumers and market players alike.

2.3. A necessary mastery of technical and regulatory constraints

When accomplishing its mission, a DSO faces many technical and regulatory constraints. Given the particular fact that investments in this field are made over periods of more than 30 years, these constraints are fundamental to the design and operation of distribution systems.

The DSO must overcome two major operational limitations: voltage fluctuations and power spikes.

Grid operators have to follow many constraints concerning power quality. Transmission system operators (TSOs) are responsible for signal frequency, which must not deviate by more than 0.5 Hz from the 50 Hz nominal frequency. For stability reasons, frequency is actually maintained within 0.05 Hz. However, voltage is a responsibility for the DSO, and must be maintained within a 5% error margin in MV networks and 10% in LV networks.

Managing voltage fluctuations is crucial, as in the absence of safeguarding action taken by the operators, significant deterioration of tension may cause a generalized network collapse. Fluctuations may also have direct impact on the equipment: during power surges, they cause deterioration and reduce lifespan, leading to increased electricity loss during under-voltage.

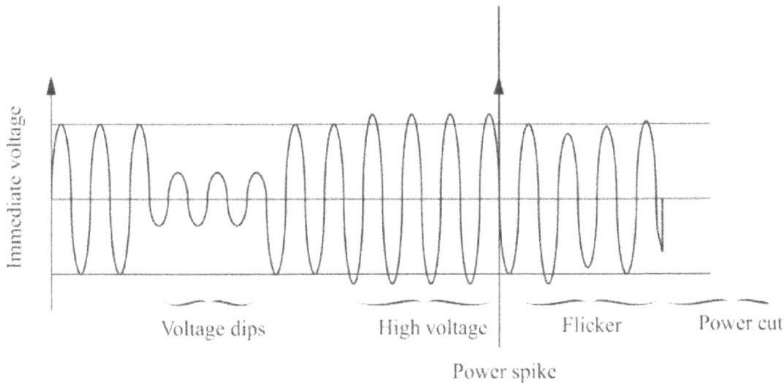

Figure 2.1. *Voltage fluctuations in detail*[1]

A DSO must also be able to accommodate the overall power consumption of all customers connected to the network. Consumption is particularly irregular and depends on the time of day, as well as weather conditions (extreme heat or cold). Other than the ability to forecast daily power consumption variations, the DSO needs to be able to anticipate the evolution of the peak power consumption. *In conclusion, the DSO should size the network in order to*

1 Source: CRE – http://www.cre.fr/reseaux/reseaux-publics-d-electricite/qualite-de-l-electricite.

guarantee sustainable operation during these critical periods (see the graphs below).

Interference	Definition
Voltage dips	Brief dip in voltage RMS value – a few milliseconds to a few seconds
Power spike	Brief or very brief spike in either RMS or instantaneous voltage value
Flicker	Rapid and relatively weak fluctuations of voltage RMS value
Under-voltage	Voltage RMS value (over a few seconds or minutes) consistently below the allowed minimum – typically 90% of nominal voltage
Power surge	Voltage RMS value is consistently above the allowed maximum – typically 110% of nominal voltage

Table 2.1. *Different types of voltage perturbations*

On top of the technical constraints, the DSO also needs to take into account the gradual evolution of regulations related to power supply quality and reliability.

Market liberalization and the unbundling of incumbent suppliers were systematically accompanied by the creation of a regulatory body. Other than supervising the market unbundling process and the activity of the actors leading it, every regulation body also provides mechanisms that ensure investment efficiency and increase of quality for all consumers of the network. Incentive regulation mechanisms are generally implemented as key performance indicators (KPIs) and result-based merit systems.

For this reason, many countries defined indicators to measure the continuity of service, generally, over a period of one year. Two indicators are directly related to customer service quality:

– System Average Interruption Duration Index, or SAIDI, the cumulated average duration of power outages per customer;

– System Average Interruption Frequency Index, or SAIFI, average frequency of prolonged outages (often > 3 min) per customer.

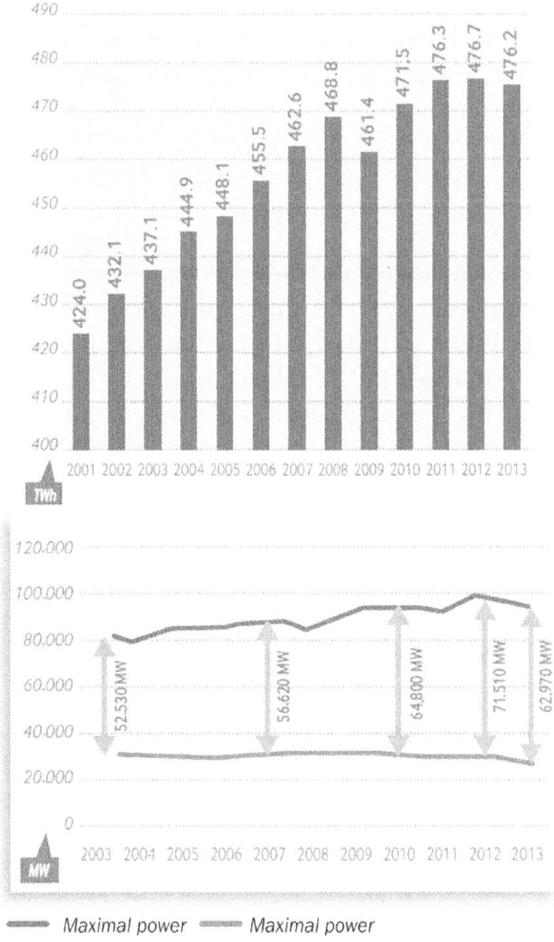

Figure 2.2. *Evolution of total consumption in France (in TWh) and annual consumption peaks (MW)*[2]*. For a color version of the figure, see www.iste.co.uk / boillot / smartgrids.zip*

2 Source: RTE – *bilan électrique* 2013.

Electricity is considered as an "essential" need, and customers are progressively less favorably disposed toward low-power supply quality. Beyond levels of quality habitual to the customer, low quality supply can have an impact on the economy of the territory. Enterprises in the digital, health, transportation and continuous industrial process sectors are particularly affected by power supply failures.

A DSO has to make use of these quality indicators to determine the volume of investments necessary to reduce failures and their economic impact.

Other than executing normal operation and maintenance (O&M) processes, distribution network operation also requires the ability to anticipate and overcome technical and regulatory constraints. DSOs make it possible by constructing an "investment master plan", predicting network evolution over the upcoming period of 5–10 years, and designing an industrial policy governing the choice of materials, technologies and even first-tier business partners.

Contemporary DSOs are currently at a turning point. Antiquating networks, in combination with the rapid integration of an increasing number of RES, and new power uses have the potential to greatly raise risk of network failure. *Hence, the use of advanced smart grid technologies must be systematized in order to control constraints and guarantee supply quality, while limiting investments into network reinforcement.*

2.4. Generalities of network design

Several key elements make up distribution networks: substations, which provide an interface between two differently voltaged networks; overhead or underground lines that carry power to the area; switching devices that manage operation schemes as well as failures; adjusting elements, such as condensers; and finally, sensors, digital equipment and software for monitoring and control. This section aims to introduce those elements to the readers, and encourage further exploration through more detailed articles and literature[3].

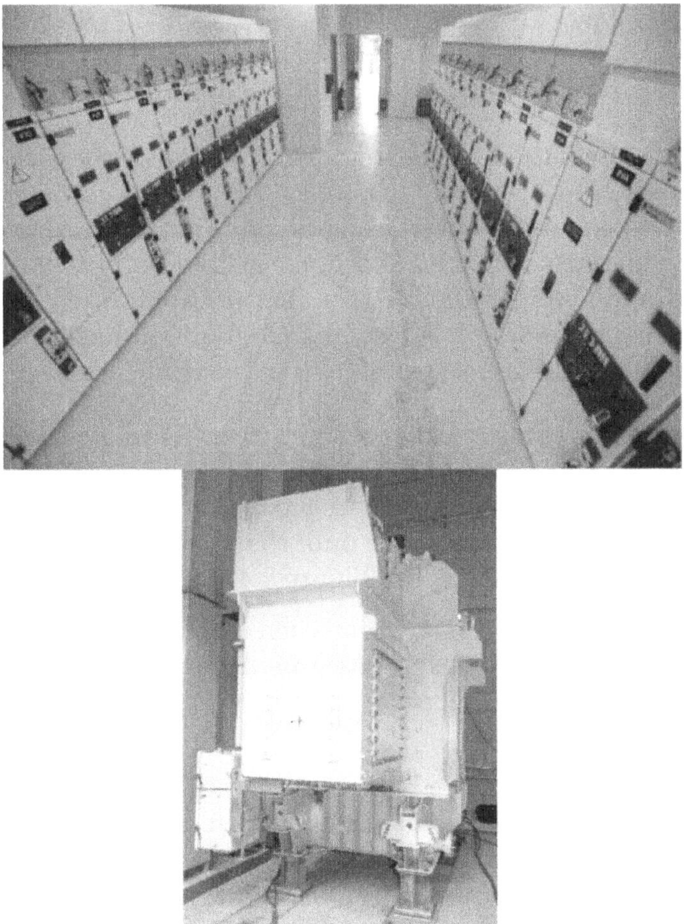

Figure 2.3. *Example of a substation: digital control equipment and transformer*

3 For examples see [HAD 11, HAD 13, DOU 14a, DOU 14b, PER 84].

2.4.1. Energy transformers

HV-to-MV stations, known as primary substations, are the link between transmission and distribution networks. Besides providing power transformation, primary substations may also be equipped with on-load tap changers, which allow 1% step voltage regulation. This function decreases voltage fluctuations in the downstream MV network, thus remaining below regulatory limits.

The distribution network is subdivided into two parts, the MV and the LV. MV/LV stations, also known as secondary substations, are the interface between MV and LV networks. In the absence of decentralized generation, in order to limit losses and voltage falls, distribution networks are operated under higher voltage. Hence, distribution stations are used to bring MV as close as possible to the consumer areas.

There are several ways in which a transmission grid can supply a distribution network. The choice of architecture depends on some techno-economical parameters, the required quality of service and the need for redundancy and/or power balancing (see Figure 2.4):

– *simple branch:* a transmission substation supplies a series of secondary substations from a single line; there is no safeguard in case of a power surge or perturbation upstream the substation;

– *"lead-through":* two primary substations supply a series of secondary substations, their interface is managed by switches; during a power-surge or some other upstream issue, the switches can be used to alter the configuration and to maintain supply quality to the downstream secondary substations;

– *double branch:* a transmission substation supplies two parallel power lines to a series of secondary substations; this allows more flexibility by distributing workload between the lines.

Secondary substations can be pole-mounted, or located over- or under-ground.

Other than transformation, secondary substations can also provide protection devices, energy meters, voltage and current sensors, control-command equipment, etc.

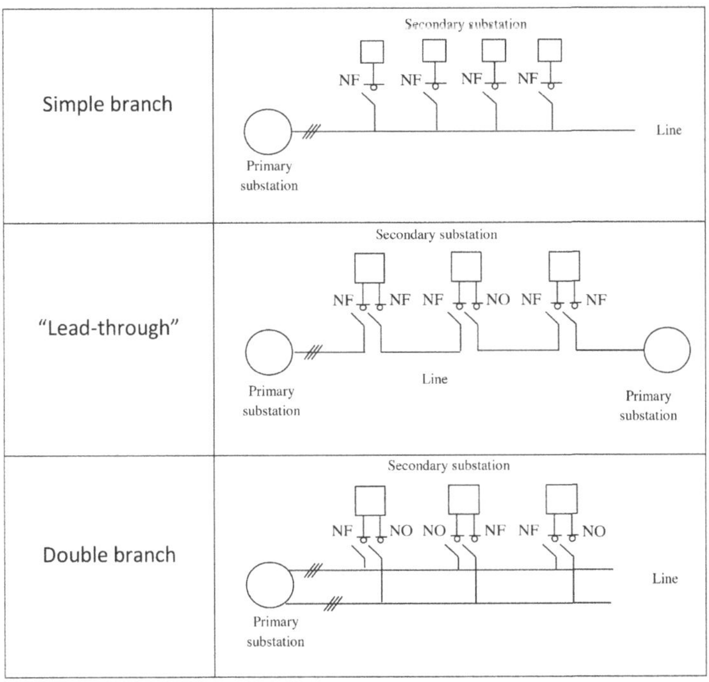

Figure 2.4. *HV/MV system supply [HAD 11]*

2.4.2. *Wiring and architectures*

Both MV and LV system cables can be mounted onto a pole or set underground. The underground solution is generally used in urban areas, as well as rural areas where weather conditions present an outage risk. The aerial solution is generally deployed in rural areas. In comparison with the underground approach, this solution is easier to maintain (detecting defects and changing cables) and is more affordable. The types of conductor (lead, aluminum, etc.) used here depend on the desired technical and economical properties.

Distribution networks generally favor an arborescent structure

This solution allows us to distribute electrical energy progressively, while scaling wire diameter and equipment size to match consumer demand. This is somewhat analogous to the circulatory system, composed of arteries, veins and capillaries.

To reinforce network robustness, "looping" architectures are created by integrating switches and defining optimal operation schemes for different situations. Such loops are positioned between neighboring substations or feeders.

In certain situations, notably dense urban areas, or areas with sensitive industries, distribution networks can be operated with a "lead-through" connection in a grid or mesh architecture, to guarantee a superior level of service:

– the mesh structure is composed of loops driven by a substation or other loops due to a dedicated link;

– the grid is more complex in design and operation than the mesh structure.

In both cases, switches are integrated into the network to configure different operation routes, depending on the situations and the level of service required. Nonetheless, these two architectures remain expensive and their operation is quite complex.

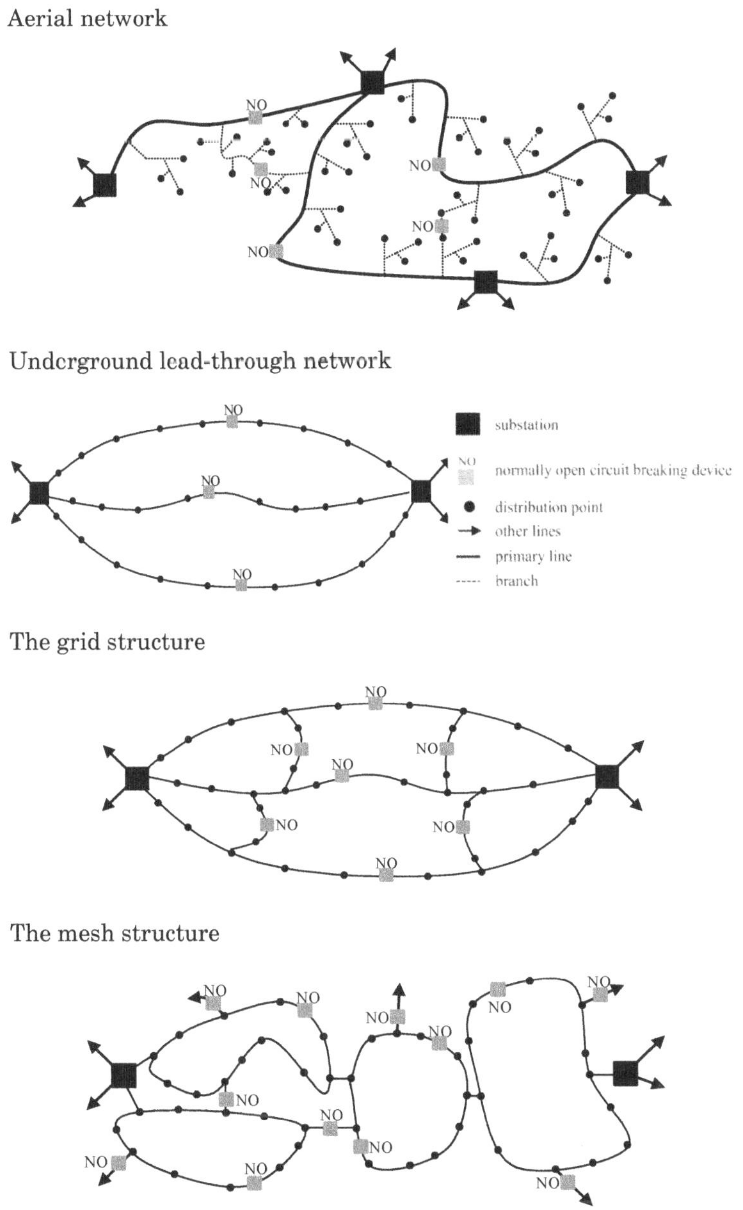

Figure 2.5. *Examples of system structure (from [HAD 12])*

2.4.3. *Safeguard devices*

Distribution networks are prone to several types of faults that require the presence of devices for localizing them and providing protection:

– single-phase fault: short circuiting between the phase and ground (majority of faults, above 70% of incidents);

– two-phase fault: short circuit between two phases or two phases and the ground;

– three-phase fault: short circuit between three phases.

To combat these situations, the DSO must provide a protection plan to preserve the security of property and people, prevent equipment destruction and guarantee continuity of supply.

The protection plan depends on neutral point treatment, i.e. the way the neutral point is connected to ground. For instance, grounding generates higher tension, which allows the creation of amperometric protection from the current circulating through the ground.

Naturally, we must first be able to differentiate this from a ground current caused by an unbalance between the phases, which can reach up to 40% of the current in some parts of the system. The problem arises from the fact that fault current decreases in a hyperbolic manner over distance from the primary substation. Over long lines, the unbalance current may mask a ground fault current.

If some faults rapidly cease naturally on their own (in under 100 ms), others need to be eliminated by fault safeguarding devices, such as fuses, reclosers and switches/breakers:

– *fuses*: fuses melt more or less quickly in response to the applied current, following a predetermined fusion curve; the configuration should have the fuse melt as close as possible to a fault. This device is inconvenient as it needs to be localized and replaced after fusion, although this solution remains cheaper than fully electronic solutions;

– *reclosers:* reclosers are circuit breakers with multiple breaking points distributed over the main feeders of the distribution network. These may be associated with fuses. Breakers open and close automatically in the presence of a fault, and during a predetermined cycle to filter out transient faults and eliminate permanent faults. Every breaker features several fusion curves. If a fault persists after the first opening/closing cycle, the recloser automatically adjusts its configuration. This operation is repeated until the fault is purged or a fuse on the line melts. If no fuse melts, the recloser will remain open and the fault will have to be localized between it and the closest fuse;

– *switches/circuit breakers*: these are placed at the beginning of branches and upstream of the reclosers. They replace fuses by counting the number of occurrences of fault current. Once a certain predefined value is reached, the breaker opens, thus preventing the fault from reappearing once the recloser closes. Compared to fuses, breakers present the advantage of being remote-controllable, which allows, once the problem with the line has been resolved, to put the entire system back online without human intervention.

In some cases, the intervention of an O&M team is necessary.

2.4.4. *Sensors, digital equipment and software*

The development of distribution networks always favors technologies that are simple, reliable, robust and economically viable. By contrast, evolution of network operation favors *improving system monitoring and control.*

Hence, information "capturing" devices and software has already been integrated into the network for several years:

– *smart meters* set up at the endpoint customers' homes; these usually measure power consumption and monitor the installation status at any time;

– *data concentrators (DCs)* that transmit the above information at regular intervals to a central system (metering system and network supervision);

– *voltage and current sensors* installed on MV lines and in secondary substations;

– *information exchange and operating devices (IEODs),* set up at generation facilities supplying the MV network, which measure voltage, active and reactive power, and exchange information for network connection/disconnection;

– *technical exchange platforms*, exchanging with aggregators that on the one hand keep track of facilities participating in flexibility programs and on the other hand receive provisional programs that reduce consumption and/or generation peaks.

The collected information is processed by various software for:

– *supervision and command* (Supervisory Control and Data Acquisition (SCADA)/Distribution Management System (DMS)), to analyze the state of the system, identify and localize faults, and carry out recovery actions when necessary;

– *forecast management* of consumption and generation based on past usage data collected by the smart meters or IEOPs, and of forecast programs by the flexibility aggregators, which allow us to estimate future congestions and constraints;

– *asset management*, to evaluate at a longer term maintenance and modernization needs based on historical data (power quality, incidents, increases in peak consumption and/or generation, etc.).

Finally, special equipment is also deployed for transmitting commands issued by the dispatching centers or at the suppliers' request:

– *remote-controlled operation devices*, which allow opening and closing breakers/switches (either to optimize the operation scheme or reconfigure the network during disruptions);

– *smart meters* that transmit pricing signals (e.g. peak/off-peak signals) and contribute to driving downstream devices;

– IEODs can be used in certain cases for decoupling requests issued by the DSO or switching to a special operational regime before performing live works.

2.4.5. *The importance of telecommunication for operating the distribution networks*

For the last 30 years, DSOs have been looking for improved performance of the networks in terms of quality and security of supply. The automation of network operation (mainly on MV) has been the key of their strategy based on SCADA centralized infrastructure: dispatching center, remote sensors, remote-controlled devices, remote-controlled primary and secondary substations. With this model, communication infrastructures have always had a major role to ensure communication between the different components of the distribution grid. When evaluating communications platforms, DSOs are looking for a solution that:

– provides cost-effective system architecture, with a strong industry support;

– is standard-based to ensure interoperability;

– provides robust, long range, two-way link;

– includes security mechanisms.

In order to ensure the best quality of services, DSOs have traditionally used a heterogeneous set of technologies: wireless radio links, GSM, dial-up leased lines and direct serial or modem connections to meet communication requirements. There is no doubt that smart grids will continue to exploit multiple types of communication from fiber optics to wireless and wire line.

Among the wire line alternatives, there is power line communication (PLC), a technology providing a communication path under the direct and complete control of the DSOs. It is a proven technology (almost 100 years) that has been evolving progressively for voice from the 1920s and data communication purposes from 1950s:

– 1950s: first applications of one-way communication for public lighting management (10 Hz);

– 1970s: very low-speed communication (20 bits/s) and still one-way communication for simple control of devices and appliance (5–500 kHz bandwidth);

– 1980s: narrowband communication (2.4 Kb/s) and two-way communication for command and control applications;

– 2000s: broadband above 2 Mhz for multimedia consumer applications.

The past two decades have also permitted to make PLC "*an excellent and mature technology*"[4] that can support a wide variety of applications for DSO and also within consumer home. In particular, since the beginning of 2012,

4 From [GAL 11].

G3-PLC technology has been homologated by the International Telecommunication Union (ITU) as a new standard[5] for narrow band PLC, which delivers a highly reliable, high speed, secure two-way communications. *Thus, G3-PLC is responding to the main concerns of DSOs and their network operations (see section 4.5).*

2.5. The factors that differentiate network architecture[6]

Although electrical networks developed worldwide present some obvious similarities, they also retain notable differences between them, which are not always easy to trace back to their origin.

For transmission grids, the existence of a globalized and powerful prestandardization body, the *Conférence International des Grands Réseaux d'Electricité* (CIGRE), contributes to normalizing choices. Moreover, the equipment market, relatively limited in volume and expensive to develop, is controlled by a handful of enterprises, which contribute further to the uniformity of choices.

For distribution networks, the situation is a little different. The networks are larger in terms of area (kilometers) and the amount of equipment. Suppliers are also more numerous, as enterprises operating the distribution network are more distributed (e.g. 3,000 DSOs in the USA and 2,500 in the European Union). Frequently, the market remains regional, and there is no powerful consultative body like the CIGRE for the transmission grids.

5 ITU-T G.9903.
6 Section based on [DOU 14b].

Nevertheless, it remains possible to identify some of the factors that explain the differences and which can structure the network configuration. We identify five such causes:

– the voltage levels;

– the MV neutral point treatment and two-, three- or four-wire distribution network;

– the balance between automation, redundancy and reliability;

– the density and layout of the service area;

– the variation in building design.

2.5.1. *Voltage levels*

This factor can be subdivided into two parts:

– the amount of voltage levels;

– the value of these levels.

2.5.1.1. *The amount of voltage levels*

At the level of distribution, the amount of voltage levels depends on the distance between the centralized generation facilities and concentrated consumption points. In interconnected medium-sized systems, 400 kV would be the highest retained voltage level. For larger systems, we could have 500 kV, 700 kV and even 1 MV.

220 kV is rather systematically used as intermediate voltage level for the transition between transmission and distribution.

Within this, two arrangements exist:

– 220/110 kV transformation, or more often, 35/10 kV;

– direct 220/20 kV transformation.

Countries that choose 10 and 35 kV, or 6 and 35 kV want to have 35 kV voltage available in rural areas and 6 or 10 kV voltage in the urban areas. Indeed, in the rural areas, long distances require relatively high voltage in order to reduce voltage drops. In the urban areas, areas with reasonable density, such as those in 1950s–1960s, during the initial electrification, 6 or 10 kV presents the advantage of being inexpensive and easy to set up, without carrying the constraint of voltage drops.

Countries that have to reconcile urban and rural areas often choose triple winding, 110/35/10 kV HV to MV transformers, to limit the amount of equipment in secondary substations, for example.

Comparative analyses of these situations leads to the following conclusion: a high count of voltage levels raises the overall cost of the network, as primary substations are an important cost factor. Furthermore, a large amount of voltage levels is generally followed by a more important cumulative length of the network.

Thus, we obtain the 220/20 kV arrangement in France, compared to the 220/110/35/10 kV arrangement in China.

2.5.1.2. *The value of the voltage levels*

For MV networks, a weak distribution voltage lowers construction costs, given that service area density remains low. It is also easier to implement.

A higher distribution voltage lengthens the feeder lines and is better adapted to low-density rural areas, although quality remains limited, as there is a close correlation between quality and network length.

We systematically found that countries that chose a low level of MV value in the urban areas currently find themselves at an impasse when charge density increases.

Certain countries combine the traditional 6 or 10 kV network with an underground 20 kV network for new areas with high density (Japan). Others maintain preexisting voltage levels at the price of choosing very high sections (China).

The change of voltage level that began in France in the 1960s is an important but difficult decision. It requires an increased capacity for foresight.

Operating voltage (kV)	Country	Rated voltage (kV)	Standard
6	Japan	7.2	CEI
10–12	UK, Germany and China	12	CEI
13.8	USA and Australia	15	ANSI
15	Deprecated	17.5	CEI
20	France, Italy and Spain	24	CEI
25	USA	27	ANSI
33	Turkey	36	CEI

Table 2.2. *Examples of voltage level in different countries wordwide*

For LV networks, two standards remain: 230 and 110–120 V. Technological constraints (security and isolation) are similar; 230 V presents undeniable advantages: reduced sections, less heating and smaller equipment. The reason for maintaining two international standards is historical. The fact that the vast majority (at least in value) of electronic equipment accepts both 110–120 V and 230 V, 50 and 60 Hz, facilitates the movement of goods manufactured worldwide and mitigates the inherent inconvenience of having two standards.

2.5.2. *The neutral point treatment in MV networks*

It may seem surprising at first that we place this sort of choice so high in the differentiation criterions between networks, given its highly technical character.

As a matter of fact, the choice of neutral point treatment entails many other choices down the line.

Neutral point treatment defines the maximum level of a ground fault current in the network, and by extent, the fault detection system, thus profoundly influencing the level of quality.

Approach	Characteristics
Neutral grounded via impedance	Simple protection, but risk of power surge
Isolated neutral	Difficult fault localization
Directly grounded neutral	Fault current is very high and risk of damage
Compensated neutral	High level of security, but complex protections

Table 2.3. *The principal characteristics of different MV neutral points treatments*

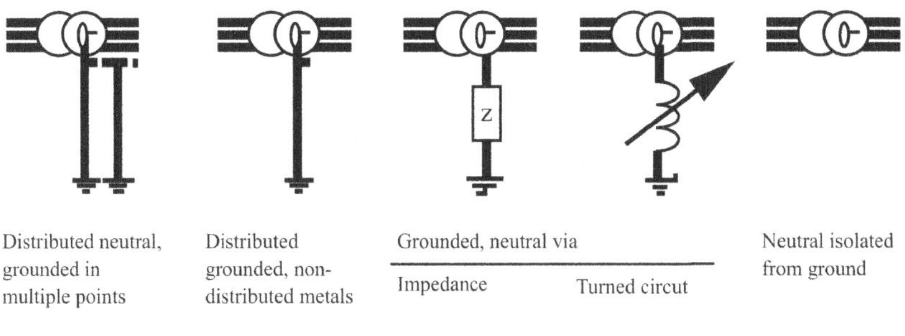

Figure 2.6. *Different schemes of neutral point treatment*

Networks using four cables with a distributed, directly grounded neutral exist in North America, parts of South America, Australia and generally in countries under the influence of the USA. Given high fault voltage, it is possible to protect the network with a fuse. *However, this rudimentary solution delayed automation of the networks.* In these networks, the neutral cable is grounded in multiple points, consequently excluding differential type protection.

In other areas, networks are three-phased, so the *neutral is non-distributed.*

Several European countries opted quite early for a *compensated neutral point treatment*, which allows for operation with a maintained weak ground fault current, as long as the voltage at the fault point is near-null (this is the case in Germany, Hungary and Poland). *In this configuration, a fallen wire is difficult to detect.*

Isolated neutral is used in Spain, Sweden, Norway, Italy and partially in China.

Asian countries often adopted the compensated neutral approach, despite the mediocre ground fault detection it offers, which leads to security issues and non-negligible fault-induced damages (for example, parts of China and Russia).

Neutral via *impedance* is used in Germany and France.

The neutral grounded directly at the primary substation is used in the UK.

Several countries are changing neutral point treatment:

– sometimes for security reasons (moving away from operation under maintained ground-leakage);

– sometimes to accompany changes in network configuration (increasing rate of underground installations);

– sometimes for issues with detecting ground faults.

2.5.3. *The balance between automation, redundancy and reliability*

This is another matter of differentiation.

The intended level of quality (implicitly stated or not) can be globally obtained in two ways:

– by using very reliable equipment, thus offering a reduced level of failure, or;

– by ensuring high redundancy to provide secondary, even tertiary, supply in case the primary one is unavailable.

Over the past few years, the achievable level of reliability has progressed a lot. Various conclusions can be drawn from this progress.

Certain countries (Russia and China) remain on high levels of redundancy. Others (France and Italy), reduce redundancy in line with the increase in reliability. This affects the substation schemes (double or single bus panels in HV and/or MV networks) and MV architecture: simple, double or triple supply.

Furthermore, the level of automation (and/or remote control) of the network varies a lot from one country to the next, depending on the architecture used. Indeed, certain architectures lend themselves better to automation, while others a lot less.

MV electrical networks protected by fuses are difficult to automate. Simple architectures are generally easier to automate.

2.5.4. *The density and layout of the serviced area*

These elements are key for structuring the LV network and, in consequence, for the type of secondary substations used.

Dense areas justify a developed LV network and secondary substations equipped with high-power transformers.

Areas that are not as dense call for low-power secondary substations supplying isolated houses. The US model represents an extreme situation, where a pole-mounted transformer is dedicated to virtually only one to five houses, which, on a side-note, could allow single-phase LV distribution (or even single-phase MV end-point distribution).

The main layout of the residential area directs the standardization of the network; hence, in the US, LV networks well adapted for rural areas can be found even in cities. Consequentially, these standards developed in countries under US influence, as well as in Japan.

It is also undeniable that the structure of the residential area had an effect on the initial balance between overhead and underground lines. This initial balance also established local standards that are applied systematically, which explains the high variations that are found today in the aerial and underground balance. In Belgium and Holland, the MV network is practically entirely underground; in fact a single construction standard exists, and it is accounted for in equal parts by the history and structure of the residential area.

France developed two different standards, one for the urban (substation box connected to the network by lead-through or double branch) and another for the rural (H61 pole-mounted MV substation).

2.5.5. *The variation in building design*

Globally, for the aerial, even different technical approaches are close in cost. By contrast, the construction choices for underground lines differ greatly by country.

This makes important impacts on the cost, and thus on the aerial/underground balance, in the countries where aerial lines are favored even in urban areas.

There are three types of architectures for underground lines:

– buried cable: the least expensive solution;

– ducted cable: sometimes gets used by various underground dwellers;

– tunneled cable: certainly the most expensive solution, requiring extensive allowances.

This choice depends on the internal rules of electricity companies and external constraints, notably in terms of safety. For example, in Germany, burying cables directly into the ground is practically impossible for security reasons.

2.6. Network safety and planning

The quality of electricity is synonymous to the voltage quality. It has two components: supply continuity and voltage consistence.

Supply continuity: depending on the duration of interruptions, we distinguish microfailures (<1 s), due, for example, to the reclosing of a breaker following a transient fault, brief failures (1 s– 60 s), due to the work of the protections, such as slow reclosing, prolonged failures (>1 min), due to incidents or scheduled work.

Voltage consistence: measured by variations in amplitude and frequency, and waveform distortions. The problems can be: over-voltage, under-voltage, flickering, noise, transient, harmonic distortion and frequency variations.

To reduce the probability, frequency and duration of power interruptions and to diminish their effects, it is necessary to invest in network reinforcement. Criteria were invented to solve this problem created by the compromise between economic constraints and reliability.

In the beginning, the criteria were deterministic: planning out the installed generation capacity in accordance to the expected demand, a spinning reserve equal to the expected demand plus a reserve equal to the greatest generation unit, system planning:

– criterion n-1: the system must continue to function within acceptable boundaries when a major element has been lost;

– criterion n-2: ditto for loss of two major elements.

This method is inconvenient as it does not take into account the stochastic behaviors of the system, of the demand or of component failure.

Typical probabilistic aspects in a power grid are: the rate of generation system failure depends on the type and size of the unit, the rate of line failure (including associated equipment) depends on its architecture, length and location.

Probabilistic analysis becomes an integrated element in the decision process that concerns electric systems, but requires a complete understanding of the system. This is a tool that allows the analysts to transform their knowledge of the system into anticipating its probable future behavior.

2.6.1. *Development of distribution networks*

Networks must be developed for five essential reasons:

– ensure that new customers are served or that new producers are connected at the distribution level;

– reinforce the existing network to account for increase in loads;

– improve supply quality in accordance with customer expectations, in terms of supply continuity (availability) and consistency of voltage;

– identify weaknesses in the system;

– assess system performance under safety constraints.

Studies allow us to match the required power availability to the network that needs to be implemented. The designers and planners identify the facilities (generators, lines, etc.) necessary to satisfy the safety criteria for future operating conditions. The planning in essence consists of finding an economical optimum within the given constraints. Consequently, it lends itself easily to techno-economical calculations *that aim to develop the network at the lowest possible cost*, while maintaining an acceptable level of service quality. Planning requires evaluating the costs of investment, operation and failure (interruption and repair costs).

2.6.2. *Operating distribution networks*

At the level of exploitation, the aim of the studies is to identify the limits of network operation within which the reliability criteria are met. Network exploitation consists of carrying out operations that yield the best supply quality to the customers from the existing network.

Network operation involves configuring the network in order to ensure optimal energy flow or to re-establish the transmission when the failure of a part of the network renders it unresponsive. It includes actions made remotely and locally at the site. The remote operation system plays an important role in operational performance.

2.6.3. *Studies in operational safety*

The aim of studies in operation safety is to provide assistance in establishing and executing various operation policies in order to avoid severe disruptions in the system. Here, the difference from system development lies in the nature of the decisions made. In the long term, planners identify means necessary to satisfying safety criteria for the expected operational conditions. The decision is whether to reinforce the network or not.

In contrast, operators identify the limits of system operation, beyond which reliability criteria are no longer satisfied. The decision is whether to take the measures to modify conditions of operation or not.

2.6.4. *Monte Carlo method*

The Monte Carlo method[7] is a stochastic method that can be used by DSOs to simulate the impact of distributed generation on a rural-type secondary substation feeder. Ultimately, this method generates safety indices for every affected customer on the substation feeder. It is defined as a model-resolution technique that uses random or pseudo-random numbers. The pseudo-random numbers are

7 This method is described in Chapter 8 of [HAD 11].

generated by applying deterministic algorithms known as pseudo-random number generators.

2.6.5. *Some results from applying the Monte Carlo method*

Rural feeders have a more arborescent structure than urban feeders (i.e. they have more secondary arteries): the goal is to study the impact of distributed power generation (DG) according to its distance from a major artery.

The Monte Carlo simulation obtains safety indices for each customer. In order to improve computation efficiency, it is possible to aggregate these customers into groups of a selected size. It is also possible to input average network indices in terms of interruption frequency (SAIFI), unavailability (Average Service Unavailability Index (ASUI)) and undistributed energy (Average Energy Not Supplied (AENS)).

Emphasis has been put on the current changes in these networks and the new challenges that are integrated in the safety studies. With the rise of the electricity market deregulation politics and the connection of new means of generation at distribution level, the networks are in the midst of undergoing changes. *The methods of distribution network safety analysis must to be adapted to these changes. One of the current trends is to apply methods previously used for transmission grids to the distribution level.*

In this respect, the Monte Carlo method is very powerful for modeling and resolving complex systems. Its main disadvantage is the amount of time needed for its calculations. However, by its nature, the method lends itself well to parallel processing and several other methods that can be used in order to reduce the resolution time. Moreover, this method gives safety indices for every point in the network and allows us to avoid establishing a model for every client.

It is also easy to combine this method with other equivalent analytical methods that aggregate clients and parts of the network according to a certain level of detail and computation efficiency. Unfortunately, it has major difficulties with events related to transitional regimes of the system (dynamics of generation units and network elements): their study requires simulations that are too long (in terms of computation time) to put into practice in a reliability study. However, certain aspects may be processed by making several approximations.

Today in operational safety studies, this method takes an increasingly important place and logically should become essential in the future, given the exponential increase in computing power at our disposal, whether it is used in preparing for operation or planning at a longer term.

2.7. Progressive modernization of a distribution network – the French example[8]

In each subject of differentiation, the momentum and weight of history are important. As opposed to many other countries, France challenged many established practices:

– switching from 15 to 20 kV level;

– switching neutral point treatment;

– shifting to underground MV and LV networks.

This capacity toward change is essentially due to the establishment of strong research and development (R&D) at the heart of an incumbent actor. With the development of

8 This section is based on [DOU 14a].

strong technical competence, the established operator always remained the industrial architect of its own networks.

The development of the networks over the 1950–2010 period is structured in several distinct periods, during which the pursued objectives evolved with the expectations of the French society.

We can distinguish several large successive phases:

– standardizing networks and accompanying urbanization/industrialization;

– reaching the minimal level of quality for every customer;

– targeted improvement of quality according to needs;

– progressive desensitization of networks toward climate hazards.

2.7.1. *Standardization (1950–1965) and expansion of the network (1965–1985)*

The period between 1950 and 1965 finalizes France's electrification, where the establishment of an R&D allows the incumbent actor to play the roles of normalizer and prescriber. The priority is to standardize technological choices to benefit from the scaling effect: the choice of 220 V for LV, launching the 110/220 V voltage change, choosing 20 kV MV and beginning the resorption of disparate voltages (5, 6, 10, 12, 15 and 33 kV).

From 1965, France undergoes important developments: growing urbanization, creation of industrial areas and parcels. New and inexpensive techniques are prescribed to accompany this rapid development: preassembled LV cables, synthetic underground MV cables, reinforced overhead MV lines and prefabricated cells for MV/LV substations.

In order to accompany the rapid growth of the distribution network, theoretical developments studies are carried out over the choice of voltage, new architectures and development costs (for instance, carry aerial lines for longer distances to reduce costs). Techno-economic calculations are also systematized to scrutinize all technical choices.

Hence, the 1965–1980 period is characterized by:

– emergence of IT that allowed the development of computing/network simulation software;

– progressive implementation of remote operation, replacing operators at primary substations;

– first experiments with automating the MV network.

2.7.2. *Achieving a minimal quality level for every customer*

If the incumbent operator succeeded in electrifying the territory, doing so at the smallest cost, and developing pertinent technical solutions, the quality was not always present and, most importantly, *the customers' expectations increased while the quality did not.*

Yet, one of the disadvantages of economical calculations is to always reason within a collective logic framework that does not solve individual problems. Another approach needed to be designed, as the studies showed that substation feeders were generally too long in the rural and there are not enough primary substations. How could this be rebalanced? Several storms resulted in very long power failures. How should the system be protected from the sticky snow? Should more underground lines be used? But, the underground solution remains expensive, can we not make it less so?

These questions led to the *launch of a program that aimed to offer every customer a minimal level of quality*. One of the major consequences of this launch was the construction of 650 primary substations. The objective was to reduce the length of the substation feeders, as a strong correlation was discovered between the feeder length and annual number of faults. The indicatively set goal was to aim for 35 km MV framework and a developed feeder length of 70 km.

Simultaneously, an increasing number of experiments were looking for cheaper undergrounds: trenches, subsoil plows, etc. Certain regions innovated by developing rural underground infrastructures.

At last, line works without power cuts for the customers became a rule: generalized use of live-working, provisional works to avoid power cuts and the beginning of the use of generators.

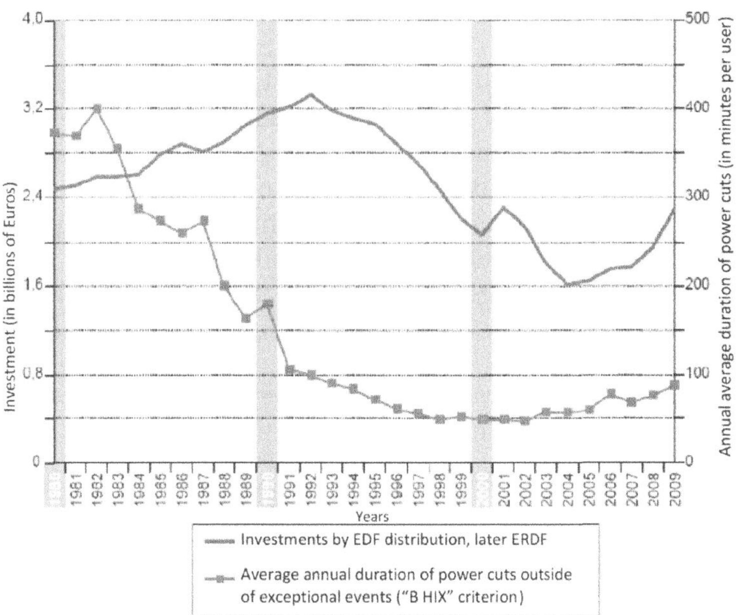

Figure 2.7. *Source: National Assembly – April 5 2011 information report on security and financing of power distribution networks (by Deputy Jean Proriol)*

Hence, in 1995, following an important mobilization of financial, technical and human resources, considerable progress was noted (results outside of exceptional conditions).

The result was close to the initial ambitious objective, with a considerable improvement of quality offered to the clients. Nevertheless, several limits were reached:

– the last constructions necessary to reach the fixed goal will be very expensive. Are they really justified?

– the selected approach allowed us to make progress in rural quality, but the urban areas were forsaken, with six outages setting a record threshold that was rarely exceeded before or since.

2.7.3. *Targeted improvement of quality according to needs*

The issues encountered by distribution network customers are diverse. Quality must be targeted before uniformly reinforcing performance, which may cost less, but does not necessarily reply to expectations. In other words, there is a clear need for personalization of quality as far as important customers are concerned (industrial, as well as small-medium enterprises (SMEs) and SMIs) and must be based on a quality standard as established in the previous section.

In particular, the quality policy after 1995 was articulated around two principal elements:

– *offering complementary services* to industrial clients with specific needs, either by providing the network or by offering an internal diagnosis;

– *actions taken near towns of over 100,000 inhabitants* in order to supply 40% of the power lost during an outage within 30 min of a total primary substation outage (following a major technological failure, a fire, etc.).

Throughout this period, technical projects multiply with the generalization of underground lines in rural areas, the launch of projects for advanced network automation (development of digital remote control, SCADA and advanced operation functions).

2.7.4. *Progressive desensitization of networks toward climate hazards*

At the end of 1999 and the beginning of 2000, two storms provoked major power outages in France: 3,500,000 customers were cutoff, the resupply of power to clients only fully established three weeks after the first outages. These two storms affected the network for the next 10 years.

Complete rehabilitation of a network takes two years, with about 10,000 km of MV lines to rebuild. Simultaneously, the following arrangements are made:

– *creation of an "immediate electricity response force" (known as FIRE in France),* a temporary organization that is capable of mobilizing within several hours teams (groups of 60 workers with their equipment) that can reach anywhere in France in case of need;

– *creation of a stock of auxiliary generators* that are permanently available;

– *reinforcement of construction standards for aerial networks* (for VHV, HV and MV lines): modification of wind reference;

– *establishment of a "climate hazard" plan* to reduce the risk of network failure in case of storms in the long run. For VHV and HV lines, the retained solution was mechanical reinforcement of the structures. For MV lines, the solution was to either switch to underground passages, reinforce aerial lines or permanently remove nearby tree branches.

Finally, important diagnostics are carried out to make note of fragile networks and establish an absorption plan: this is the "climate hazards plan" that also incorporates risks of flooding, heat waves (this section became tangible following the heat wave of 2003) and sticky snow. The continued development of underground passages in rural areas is accompanied by the development of equipment adapted to this technology: manual and remote-controlled switching substations, substations at the foot of the pole, etc.

3

Main Drivers and Functions of Advanced Smart Grids

3.1. Drivers of the evolution of distribution grids

3.1.1. *Massive integration of renewable energy sources*

Initially, distribution networks were designed to distribute electricity from the upstream to the downstream, that is, from primary substations to the end-point customers. The strong development of renewable energy source (RES) generation, distributed and intermittent (wind and photovoltaic), is at the origin of a major change to the *modus operandi* of the electric network in order to provide bidirectional flow of electricity and for the management of the increased risks of local overvoltages and congestions (see Chapter 2).

The inclusion of an increasing number of RES generation facilities therefore requires combining two major types of solution for maximizing technical and economic efficiency.

First and foremost, it is necessary to favor the integration of new RES generation according to the carrying capacity of existing networks and the consumption requirements in proximity. These parameters need to be taken into account in planning and choosing investments for the infrastructures.

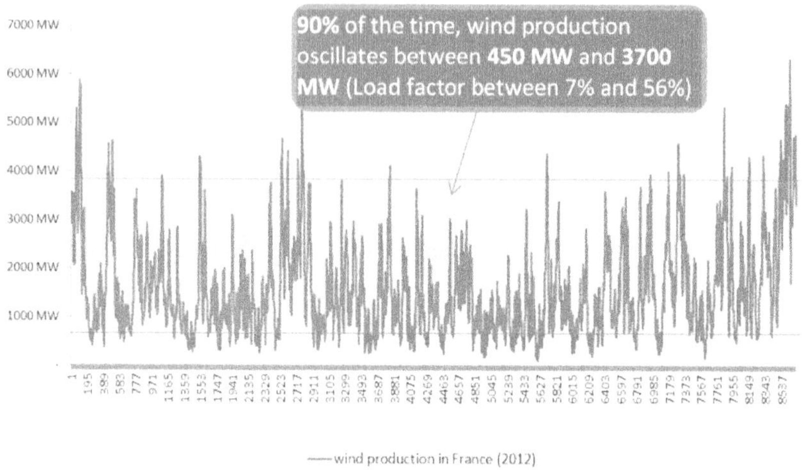

Figure 3.1. *Evolution of wind power generation in France over 1 year (source: RES Pool project presentation, supported by the ADEME, during Smart Grid Paris (June 2014))*

Once the RES facilities are connected to the network, it is necessary to set up active management of generation, consumption and network operation with advanced tools and solutions for mastering intermittence (see Figure 3.1). These tools and solutions need to contribute to reinforce the observability and the controllability of medium- and low-voltage networks:

– *measure* electrical values, specifically the RES power injected into the distribution network;

– *estimate* instantaneously the state of the electric system;

– *activate* flexibilities (modulation of active power, request for consumption, etc.), contributing to the dynamic management of constraints;

– *influence* generation and other means related to the network in order to guarantee the security and efficiency of the electric system, in particular, to manage the consequences of incidents on the network in the presence of multiple decentralized generation sources.

3.1.2. *Contribution to the development of electric vehicle and the charging infrastructures*

In France, the government predicts 2 million electric vehicles (EVs) by 2020, the most ambitious scenarios predicting between 7 and 9 million EVs by 2030.

The 2020 ambition requires installing approximately 400,000 public charging terminals and 4 million private terminals: for individuals and in supermarkets, private buildings, public carparks, etc.

Much like with the integration of RES, supporting the development of EV is based on a combination of two solutions to maximize the technical and economic efficiency: the optimization concerns the type and the location of public charging infrastructures, on the one hand, and dynamic and smart management of charging cycles, on the other hand.

As far as public charging infrastructures are concerned, it is up to the local authorities to take responsibility for its development. Nonetheless, the distribution system operators (DSOs) still need to provide as much upstream support to the communities projects as possible, in order to optimize the placement of charging terminals in accordance with the carrying capacity of existing networks, and to limit the needs for reinforcement. In fact, depending on the type of charging, the power demand may vary greatly. For example, after a journey of 120 km, a complete recharge of an EV is equivalent to the power demand of:

– a heater, if the charging takes place over 8 h;

– a building, if the charging takes place over 2 h;

– a neighborhood, if the charging takes place over 15–20 min.

Therefore, charging several vehicles in the same area during peak consumption hours can create constraints (voltage variations or overvoltage) on distribution networks that were not sized in order to be able to supply these new loads simultaneously at any point in time. Hence, these parameters need to be taken into account during the connection phase of the new public infrastructures for EV charging and in the choice of investments.

The second challenge of developing a large fleet of EVs is the dynamic management of the charging cycles and the power demand. It is preferable to coordinate the charging of vehicles with generation and network availability as much as possible, in order to maximize the service-to-cost ratio for the users, the collective and the network operators.

Advanced tools and solutions are necessary to:

– measure the periods and the duration of the charging, the quantity and the power of the electricity consumed;

– anticipate local constraints on the networks (consumption peaks, overvoltages, congestions, etc.);

– coordinate the charging cycles according to the pricing signals and/or market mechanisms (load shedding, storage, etc.).

These tools and solutions will need to be adapted in accordance with contractual mechanisms that remain to be defined between the market players. For instance, this is the case for collective charging facilities belonging to a single operator (building, office, parking, etc.) for which charging management and optimization models have not been established yet.

Figure 3.2. *ERDF showroom at Paris Grenelle*

3.1.3. *Implementation of new market mechanisms (peak shaving, capacity market, etc.)*

To limit the growth of consumption peaks and the use of carbon dioxide (CO_2) emitting generation means, different countries are studying the implementation of market mechanisms, for example peak shaving and the capacity market.

As the generally regulated players, the DSOs do not have the authority to intervene in the competitive field of these new mechanisms. Nonetheless, the DSOs remain guarantors of the network safety and the supply quality, and hence position themselves as facilitators of the implementation of these new mechanisms. In the case of peak shaving, this role consists of, for example, ensuring better characterization of the impacts of peak shaving on the distribution network. In the case of adjustment and capacity mechanisms, this role is reflected by participation throughout the three stages of their implementation: certification of *ex-ante* capacities, real-time information exchange and *ex-post* performance assessment.

Parallel to the implementation of new market mechanisms, the DSOs study the convenience of, in *the midterm, implementing levers that enable the optimization of distribution network operation. For example*:

– the capacity of the DSO to contract with market players (according to modalities that have to be defined) flexibilities for consumption and/or generation modulation to selectively reduce network constraints;

– the implementation of transmission pricing with temporal modulation in order to better reflect the cost of network constraints and hence send clearer pricing signals to the customers;

– the capacity of the DSO to operate targeted load-shedding without consulting with the customers, in order to safeguard the balance of the system.

The nature and pertinence of levers will need to be adapted in accordance with the evolution of electricity markets and the regulatory conditions, such as the coexistence of national markets with the management mechanisms at the local scale.

In the case of France, mechanisms were implemented several decades ago and additions are incoming.

This is the case with consumption shaving. The established operator optimizes the consumption/generation balance in its area by offering specific rates and services, in particular:[1]

– peak hours (PH) rate (16 h per day)/ off-peak hours (OH) rate (8 h per day, generally at night), associated with supplying certain uses (e.g. 12 million hot water tanks supplied in France);

1 http://particuliers.edf.com (gestion de mon contrat/options tarifaires).

– “EJP” (“peak day shedding” rates – 22 days per year, during the winter period) and “Tempo” (pricing with six time-seasonal ranges – the “color” (green, white or red of the rate is communicated on the eve at 20:00) offers, also associated with a supply signal, propose rates lower than the base rate, outside of extreme consumption peak periods. Opening the Tempo signal to other suppliers is currently under consideration.

This is also the case with adjustment mechanisms. Since 2003, the transmission system operators (TSO) have implemented a mechanisms that allowed it to call for the flexibilities offered by generators and consumers (directly or via load-shedding operators) in order to manage the hazards to the demand–supply balance, as well the transmission grid constraints. This mechanism anticipates a remuneration for the players.

Looking for a larger and quicker development of consumption shedding, the French legislative body provided a mechanism related to shedding, with the possibility of a double valorization in energy and power:

– *Energy valorization* is included in Article 14 of the Brottes law of 15 April 2013[2]. It allows a shedding operator to shed customers without the agreement of their supplier and institutes the principle of compensating the suppliers in accordance with the losses incurred by the shed customers. The Brottes law provides a premium paid to the shedding operators, financed by a tax.

– *The valorization of guaranteed power* is taken into account by a capacity mechanism for which the decree was published on 18 December 2012 and which now needs to be reflected in the rule set for the 2014 application for the first

2 N° 2013-313 aiming to prepare for transition towards a sober energy system with different articles on water and wind energy tariffs. Article 14 concerns load shedding and the contribution of market players to reach the energy policy objectives.

winter included by the decree (2016-2017) and the following winters.

3.1.4. *Participation in the development of new uses contributing to energy efficiency*

The improvement of energy efficiency and energy conservation are the major priorities in many countries worldwide.

Thus, for example, the European Union in 2012 set up a Directive that set an objective to lower the projected energy consumption trend by 20% by 2020. Among the energy efficiency levers, the obligation for countries to renew every year 3% of their real estate and the obligation for energy suppliers to conserve 1.5% of their annual sales volume are some key elements.

Certain Member States of the European Union also defined objectives. This is the case in France, where goals were set for the most energy-intensive sectors by the Grenelle laws. For example, in the building sector, this law sets a global objective to reduce energy consumption by 38% by 2020. To achieve it, the law also establishes a trajectory for renewing accommodations *(400,000 accommodations per year,* starting 2013) and for *the improvement of energy performance in tertiary and public buildings* (over 8 years, beginning 2012).

Even if the DSOs generally do not have quantitative objectives, they can contribute significantly to the demand side management (DSM), notably through the deployment of smart meters:

– Consumers would then be able to access detailed information on their consumptions, understand the impact of their daily usage and, in the latter case, alter their electricity consumption tendencies.

– Collectives would also have access to aggregated data (at the scale of the commune or at finer scales – neighborhood, road, block, etc.) for the implementation of their energy strategy.

3.1.5. *Urban renewal and the rise of the smart city in favor of resource optimization*

Today, although cities occupy 2% of the Earth's surface, they house 50% of the world's population, consuming 75% of the energy generated, and are at the origin of 80% of CO_2 emissions. Cities increasingly exhibit an eagerness to better manage their resources, whether that is by limiting energy and water consumption or by optimizing generation, particularly by using renewable energies.

The blossoming of new technologies makes possible the experimentation and development of new solutions for energy management and optimization at different scales, from a building or a neighborhood to a city or even a larger territory. For example, with the arrival of solutions such as smart meters, energy boxes, energy managers, Web portals and smartphone applications, accommodations and buildings become *smart homes* or *smart buildings*, whose consumption and load curves can be optimized. Other technological solutions, such as photovoltaic generation and, over a longer term, electricity storage, can contribute to reducing the energy consumption of buildings, and even render it positive (BEPOS: *bâtiment à énergie positive* – positive energy building). Finally, the integration of charging infrastructures for EVs at the level of these new territories also offers a modulation lever for energy demand.

The integration of these new objects into the distribution network to develop "sustainable cities" constitutes different challenges for the DSOs.

The DSOs have already become players in energy optimization by ensuring the panning of electric networks on the medium and long terms by taking into account city projects. *The particularity of the network is to produce economies of scale and to develop the resources of different territories.* The emergence of an optimization stage needs to take place in coordination with the DSOs and the local authority. In fact, the search for economic optimization can be detrimental to the global interest and at a higher cost: *the sum of local optimums does not necessarily correspond to the general interest.*

In addition, the emergence of these new objects requires a large volume of data related to the concerned area to be aware of the situation at the start and follow the evolution of performance in the long run: consumption and generation in the area, load on secondary substations supplying the area, topology of the network, etc. The DSO is therefore at the heart of the evolution of territories to accompany the implementation of local collective policies.

Finally, the integration of smart cities (or new objects composing them) requires changes to the business processes of the DSO in order to apprehend the different stages from planning (e.g. definition of load models and customer behavior) up to the operation and usage of the network (e.g. validation of optimization programs and requests for their modification).

3.1.6. *Integration of energy storage solutions*

Currently, two types of storage can be used with the distribution network: storage of energy as an end-point product (e.g. refrigeration or hot water tank) and electric battery storage.

As certain studies show, to this day, battery storage is not an economic solution for accommodating only the needs of medium- and low-voltage networks. Several services can provide the flexibility expected by the network operators: power plant, network reconfiguration, demand flexibility, disconnection of RES, etc.

Two conditions are necessary to respond to the needs of the electric system and bring value to the collectivity: the viability of the technology in relation to alternative solutions and the development of a *market design* appropriate for the emergence of balanced, even positive, business models. In this respect, distributed storage devices to this day do not constitute (in the context of current purchasing obligations) a profitable economic alternative to network reinforcements.

However, if the decrease in the battery acquisition costs is confirmed in the upcoming years, storage may become interesting under certain specific conditions.

Figure 3.3 presents different energy storage technologies along two axes:

– in abscissa, the ratio between the level of investment and the available power;

– in ordinance, the ratio between the investment level and available energy.

The pumped storage power station (PSPS) remains the most competitive solution for large-scale storage. Lithium-ion batteries, despite their popularity in general public devices, are not yet competitive enough to be used in energy storage.

Currently, the condition necessary for making storage profitable is that they should bring several services and address the needs of all the players of the electric system: the generator, the TSO, the DSO and the consumer.

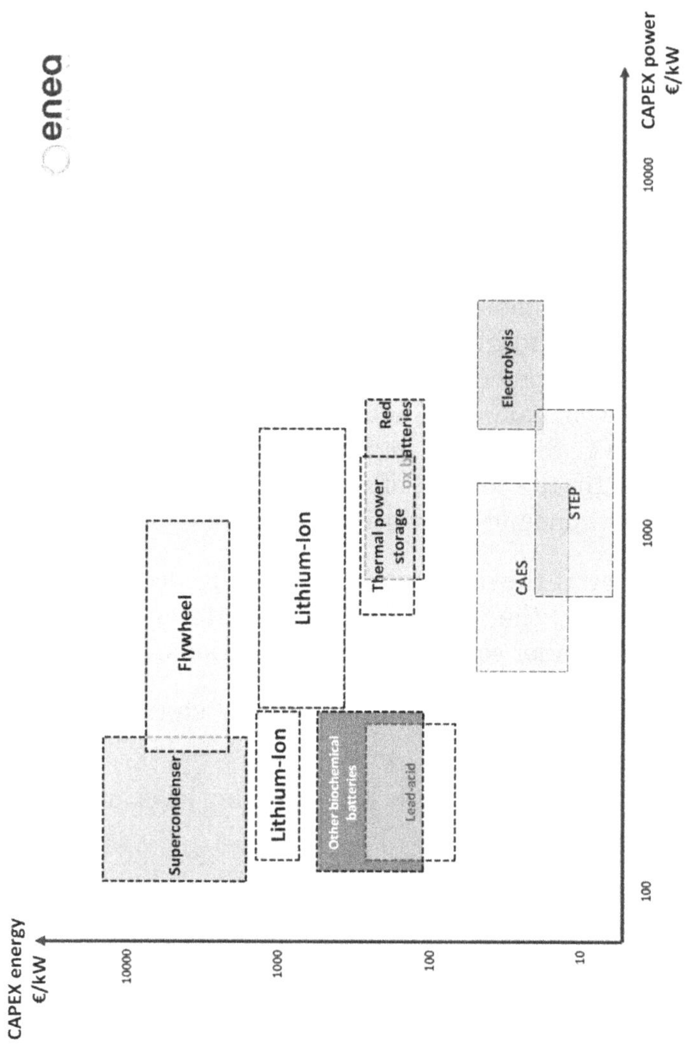

Figure 3.3. *Positioning of energy storage technologies according to their capital expenditure (CAPEX) in power and capacity (source: report of October 21, 2013 (ADEME, DGCIS and ATEE) on the potential of energy storage; graph – ENEA Consulting)*[3]

3 www.presse.ademe.fr/files/peps–rapport-public1.pdf page 183.

Indeed, from a technical point of view, electricity storage technologies can bring many services to the electric system[4]:

– *participate in arbitrages and valorize energy transfers* by optimizing charging (during low-rate periods) and discharging (during high-rate periods) cycles;

– *provide an active power reserve*, making possible the participation in regulating the frequency of the network and/or adjustment mechanism;

– *even out intermittent generation by* shaping the generation curve according to a form that is adapted to consumption and can be exchanged on markets (typically a generation block whose insertion curve is stepped);

– *absorb or inject reactive power for voltage regulation*;

– *manage transient congestions on the network*;

– *smooth end-point consumption peaks* and reduce the consumption of fossil fuels, therefore avoiding investments into new peak and extreme-peak plants.

From a theoretical point of view (by abstracting the economic pertinence of the solution and the feasibility of the implementation), electricity storage can allow network operators to:

– *defer certain investments into network reinforcement* by reducing certain consumption/generation peaks or by addressing transient consumption constraints;

– *contribute to the stability of network operation* (by bringing a solution to the eventual intermittent generation forecast or by providing system services);

4 DGCIS ADEME ATEE study on the potential of energy storage (October 2013). www.presse.ademe.fr/files/peps–rapport-public1.pdf.

– *contribute to the security and the quality* of the customers' electricity supply (especially in terms of voltage).

Just like load shedding capacities, the DSOs need to be able to master the impacts of storage infrastructures on the public distribution network (PDN) in order to guarantee the security and safety of the end-point customers' power supply. The synchronicity of storage extraction and injection with the existing users of the distribution network could, for example, invalidate the network sizing and security hypothesis.

ILLUSTRATION[5].– NiceGrid is experimenting with several types of lithium-ion batteries and planning the deployment of 1.5 MWh of batteries distributed over three distinct levels of the electricity distribution network:

– *storage at the level of the primary substation* of the Carros area, which will ensure the connection between the RTE network and the ERDF: 560 kWh/1.1 MW lithium-ion battery;

– three sets of *integrated storage devices on the low voltage (LV) network,* to manage consumption and photovoltaic generation peaks, enable islanding and manage the voltage plan: 106 kWh/33 kW lithium-ion batteries;

– several *storage facilities at residential volunteering customers' homes* to contribute to load shedding: 4 kWh/ 4.6 kW lithium-ion batteries.

5 http://www.nicegrid.fr/nice-grid-le-stockage-d-energie-10.htm.

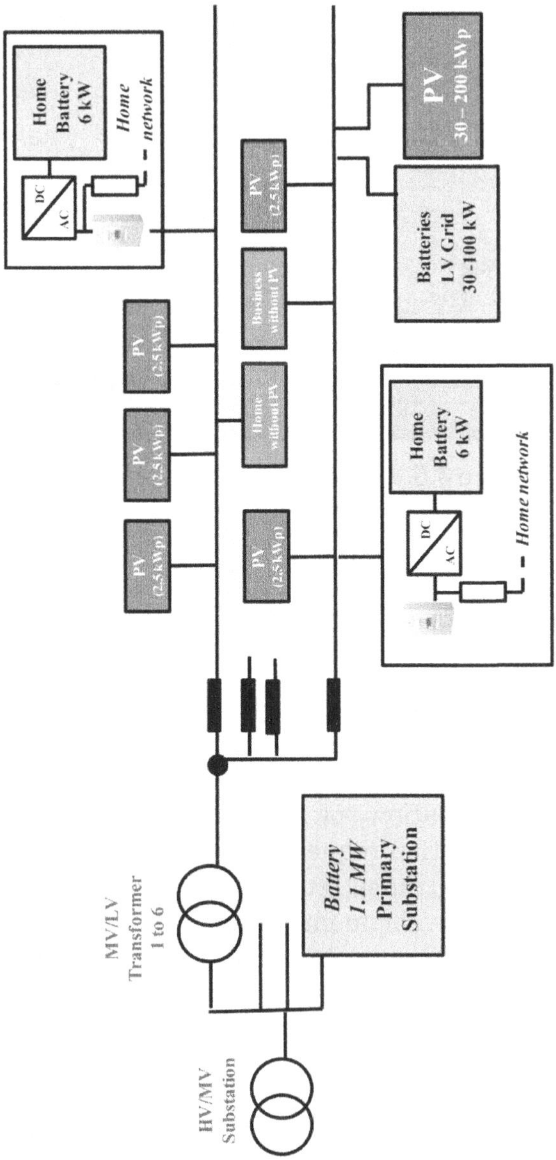

Figure 3.4. *Simplified view of electricity storage batteries integration in the NiceGrid project*

Figure 3.5. *1 MVA battery and associated power demand electronics during installation in Carros (France) press release extract – July 14, page 7*

3.2. Main functions of the advanced smart grid

3.2.1. *Toward dynamic network management by the distribution system operators*

The DSOs need to overcome the consequences of the increasing permeation of intermittent RES generation: increase in connection requests and reinforcement costs, and the reduction of transit energy. The DSOs also need to prepare for the development of new uses, in particular EVs and heat pumps, in order to simplify their insertion into the distribution network at the best cost. *The DSOs need to make their networks more "active" in order to manage dynamically the newly created constraints.*

The future model will be based on the *dynamic management of the network*. This approach implements the concept of smart grids, where technological innovations in equipment and information and communication technologies (ICT) are combined to enable more efficient operation of the distribution network.

The smart grid functions will prioritize the use of the *secondary substations,* which will become a point of privileged observability of the medium voltage (MV) and LV networks, as well as a telecommunication node between the information systems (IS) of the DSO and the downstream usages (generators and consumers). Although smart grid functions have different requirements, there is a common core for the sensors and the communication devices at the distribution substations. We anticipate the development of a modular "smart secondary substation", which will allow us to limit deployment costs. To determine the performance level of the equipment and the communication of distribution substations, it is essential to coordinate all the projects through a telecom directing scheme.

3.2.2. *Structuring the target model based on key functions*

Functional analyses were conducted by certain organizations, such as the "Task Force for Smart Grids" of the European Commission and the EDSO for Smart Grids, as well as by the large European DSOs. Advanced functions according to the current DSO model are in the process of development or deployment.

These functions will contribute in particular to six goals[6]:

– enhancing efficiency in day-to-day grid operation;

– ensuring network security, system control and quality of supply;

– improving market functioning and customer service; and

– enabling the network to integrate users with new requirements;

---

6 Source: EU Commission Task Force for Smart Grids – Expert Group 1 "functionalities of smart grids and smart meters" – December 2010.

– better planning of future network investment;

– enabling and encouraging stronger and more direct involvement of consumers in their energy usage and management.

The following paragraphs describe the advanced functions contributing to the first three goals.

Thus, the deployment of smart meters will allow *collecting and providing data issued from metering systems* (consumption, generation and supply quality). By also making use of the devices downstream of the meter, the function for *transmitting consumption data* to the clients should support the implementation of energy consumption and energy efficiency control. Moreover, the implementation of energy managers connected to smart metering should accentuate changes in consumption profiles: the control of exchanges through this interface constitutes an essential key to network development and optimization.

Metering data, as well as data from networked sensors (primary substations, MV feeders, remote control device (RCD), secondary substations), represent a *large volume of information, whose exploitation will enhance the performance of the electric system,* in particular the pertinence of investment choices (network development and quantification of customer supply quality).

In order for the "enhanced" collected data (flows, data associated with works, external information, etc.) to be used efficiently, functions, specifically IS functions, *for associating data collected from network facilities and weather reports, for creating histories,* for correcting and for security need to be implemented. The significant volume of information requires the deployment of solutions for processing this data in bulk (*big data*).

It will become possible to operate the electric network closer to its limits with the activation of different flexibility levers: load shedding activated by suppliers or aggregators, even the DSO, storage or release of energy, pricing signals sent by the suppliers or the DSO, increase in carrying capacity of decentralized generation means via dynamic adjustments to the voltage plan.

The functions for active management interface with markets and for technical validation of load shedding/tertiary generation programs will become essential for the DSO to remain capable of controlling the flow and the phenomena occurring on the network. Concurrently to the management of physical flow, the DSO will also have to adapt its tools and have the necessary information to establish corresponding financial flows through a function for *re-establishing financial flow and balance between the players of the electric system.*

Thus, controlling the network of tomorrow implies a better knowledge of the state of the distribution network both in real time and in the near future. Network optimizations can be implemented for these two conditions: dynamic management of the voltage plan to optimize the RES generation, minimization of losses, and optimization of the subscription for withdrawal from the transmission grid.

The DSO needs to be able to implement these functions dynamically, allowing the optimal management of the network of tomorrow, including decentralized generation and EVs. The acquisition and processing of the collected data (network, facilities and telecom) constitute the necessary core to the development of other functions, notably the following:

– *optimization of operation and the use of MV and LV networks* (outside of incidents), of *short-term forecast management and of network development*;

– *dynamic management of incidents,* improving reactivity to failure, therefore reducing the duration of outages;

– *network equipment monitoring functions* will provide constant information on the state of health and aging of the equipment, offering the opportunity to *optimize maintenance* and even anticipating incidents;

– *arbitrage between available levers to optimize network performance* will be necessary to activate the most economically efficient levers at a given moment.

Consumption and generation forecasts at different timescales will be necessary to drive forecast management, network studies, acquisition of losses, etc.

The equipment, information and communication infrastructures deployed will also grant control over the power consumed by the customers, decentralized generation, EV charging infrastructures and decentralized storage.

3.2.3. *Enhancing efficiency in day-to-day grid operation*

3.2.3.1. *Advanced function: supervision of medium- and low-voltage networks*

3.2.3.1.1. MV supervision

This function addresses national dispatching operators. It allows us to:

– *benefit from instantly available information about* the electrical situation of MV networks (load and power) with sensors and tools for approximating state;

– *manage alarms* generated from constraints affecting the distribution network, incidents and failure detection equipment;

– *know in advance (1 h beforehand) about short-term constraints* from generation and consumption forecast calculations.

This is a key function for guaranteeing the safety of MV network operation. The incoming data from different equipment on the electric network are analyzed and used optimally (meters, concentrators, substations, information exchange and operating devices (IEOD), RCD, etc.). In the case of an incident, the diagnosis is simplified to facilitate the rapid reconfiguration of the network.

3.2.3.1.2. LV supervision

This function is developed to give the operator a complete overview of the MV and LV electric networks. It makes it possible to find out about constraints and failures on the LV network, as well as to manage alarms. It uses sensors and tools for approximating the state.

The sensors of the advanced metering infrastructure (AMI) system are capable of giving precise indications on the state of the LV electric network (failures) or on the quality of service (with respect to voltage levels).

Other voltage and power sensors are installed in certain vulnerable distribution substations (concentration of RES or EV, voltage excursions, etc.).

The measured data are communicated to the regional dispatching through existing communication channels (AMI or other telecom systems) after an initial treatment at the local level.

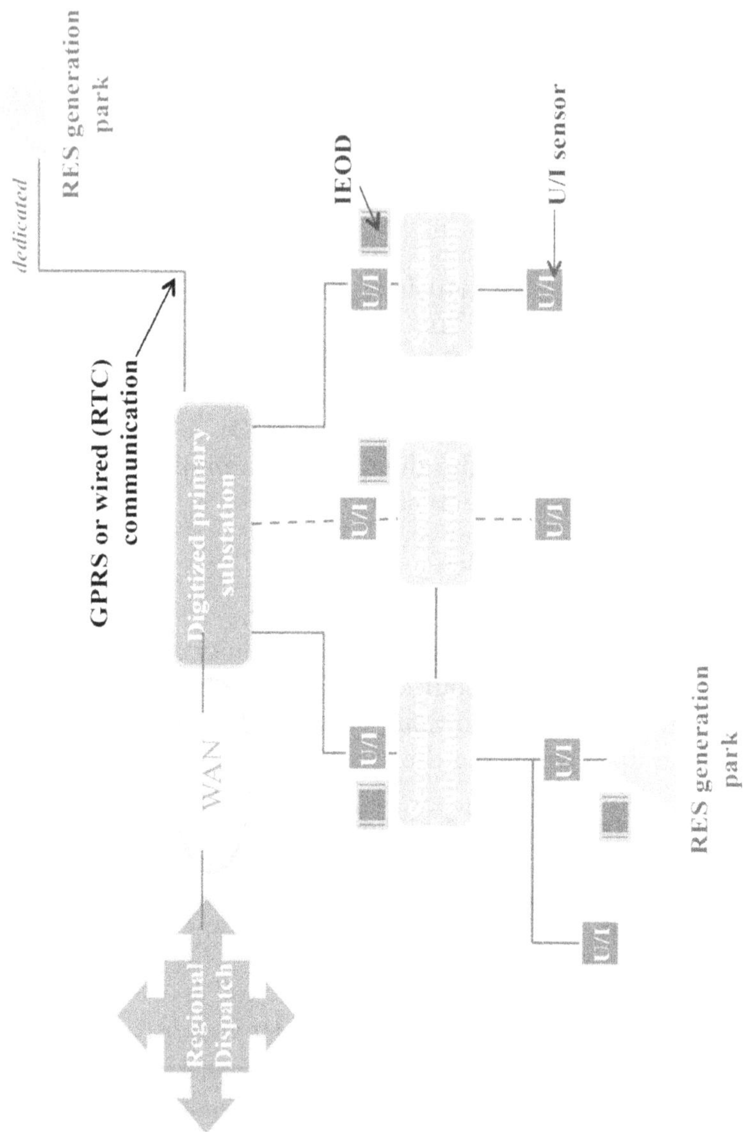

Figure 3.6. *Diagram of information transmission toward regional dispatch*

3.2.3.2. *Advanced function: dynamic management of disruptions*

This function allows us to identify parts of the MV or LV network that are affected by a failure in order to isolate them and re-establish a new operational scheme:

– *Self-healing network*: associated functions determine precisely the location of the failures on the network by using the data issued from networked equipment (concentrator on the supplied substation).

The developed automations calculate the most efficient plans to resupply the most customers in the least time. This allows many concurrent failures on the electric network to be processed simultaneously, which is not possible for a single operator. Moreover, the analysis of information, such as voltage and losses, is considered in the calculation in order to maximize the efficiency of the resupply plan.

– *LV failure detection*: the operator will know about the failures affecting the LV network through regular power line carrier (PLC) communication tests between the meters and the concentrators sharing the same LV feeder. The operator will be able to diagnose the origin of the failure: loss of a phase, neutral point rupture, etc.

– *Communication following a failure*: if the customer calls about a failure, the operator can check the customer's meter remotely and diagnose the situation; it could either be a failure related to the network or to the customer's own electric facilities.

– *Communication during supply interruption*: the operator communicates with the customer via the Internet or over the phone about planned and accidental outages, as well as their anticipated duration.

3.2.4. *Ensuring network security, system control and quality of supply*

3.2.4.1. *Advanced function: operational planning and scheduling*

This function defines optimal operational schemes for MV and LV networks (from one week to one day in advance). The data are then introduced into network operation software.

Consumption and generation forecast management will notably make use of weather forecasts. In particular, it will facilitate the more active participation in the safety of the electric system (by decreasing constraints on the network), the optimization of the management of the network and the load on the distribution network facilities, the optimization of the RES generation and the technical validation of the generation or load shedding programs suggested by the market players (suppliers, aggregators, etc.). The MV and LV network constraints are calculated from consumption and generation forecasts. This calculation also takes weather forecasts into account. The goal is to limit the constraints in order to increase the safety of the overall electric network, while accepting the RES generation.

This requires the DSO to analyze generation or load shedding requests from the generators or the aggregators to determine the strategies to follow and ensure their technical feasibility by giving its consent. These demands need to be anticipated and kept up to date.

Simulations and state approximates are used to characterize the risks related to maintaining voltage; forecast management data (generation and consumption) are supplied to a "load-flow"-type software. This enables the operator to know the overall state of the network by correctly installing a small amount of sensors into the network.

Two options are available to lower the level of constraints: network levers and flexibility levers (see Chapter 5).

3.2.4.2. *Advanced function: optimization of network operation and maintenance*

This function optimizes the operation of MV and LV networks by using the available "network" levers (voltage set-points, operational scheme modification, connected generation set-point, etc.).

The objectives pursued are the following:

– *dynamic control* of voltage to accommodate a maximum of RES generation. Several levers can be used:

- influencing consumption or generation of reactive RES energy;

- acting on capacitors (at primary substations and on the network) to compensate for voltage decreases;

- acting on the transformer load tap changers inside the primary substations (or even secondary substations if some new transformers allow it) to modify the voltage output of the transformers;

- modulating consumption and generation to modify the voltage profile along the MV and LV feeders;

– *minimization of technical losses*: computing the overall state of the network, estimating technical losses and elaborating solutions with new operational schemes;

– *recognization of abnormal LV operational schemes*: this will be possible by using the smart meter to spot abnormal schemes (e.g. employed after an incident) and to correct them;

– *participation in the security of the global electric system* by contributing to limit the constraints on the grid of the TSO.

3.2.5. *Improving market functioning and customer service*

3.2.5.1. *Advanced function: demand–response management system and interface with market*

This function ensures the coordination between the generators, the suppliers and the aggregators. They will put into place adjustment and load shedding mechanisms with their customers and transmission and distribution network operators. Moreover, the new local flexibility levers can be solicited by the DSOs. It is necessary to develop tools in order to establish certifications, guarantee data flow and ensure the transmission of commands.

Specifically, this function aims to establish a link between the customers, their connection to the network, and to keep track of their contract with the suppliers and the aggregators. It will help to determine the customer consumption volume that can be shed and to estimate the capacity of generation that can be activated.

3.2.5.2. *Advanced function: relationship with TSOs and others actors of the market*

In the interest of safeguarding system security, a framework for efficiently exchanging operational information between network operators (and between network operators and end-user customers) should be defined. The TSOs should typically rely on the DSOs to provide them with the operational information they need from end-user customers connected to distribution networks.

In addition, network operators will need more information on the planned actions of aggregators and independent power producers connected to their networks. Their participation in TSO balancing markets could create constraints in the distribution grid.

3.2.6. *European network codes*

The Third Energy Package empowers the European associations of Transmission System Operators (ENTSOE-E and ENTSO-G) to prepare the network codes establishing Europe-wide binding rules for electricity and gas markets. The European Agency for the Cooperation of Energy Regulators (ACER) provides the overarching framework for ENTSOE-E's work, as well as "reasoned opinions" on the codes. The codes then go through a process coordinated by the European Commission (EC) (an approval procedure with scrutiny) (Article 8 of Regulation 714/2009) and become legally binding.

"The network codes should be developed for cross-border networks [...] and market integration issues, without prejudice to the Member States' right to establish national codes which do not affect cross-border trade."[7]

It is important that the network codes should set safe rules for all grid users and network operators while being flexible enough to provide DSOs with some leeway to adapt the rapid changes in their networks.

According to the ERDF deputy CEO during the panel session "Market Design" at the InnoGrid2020 conference in March 2014: "together, TSOs and DSOs can find the right solutions [...] it would have been more effective and less time consuming if three years ago the European Commission would have charged TSOs and DSOs to propose the grid code needed for Europe together [...] I very much appreciate that ACER is also supporting this view [...] It will remain the DSOs' key responsibility to ensure the stability and safety of their electricity network [...] voltage control is becoming a very important issue that DSOs have to solve".

---

7 Regulation no. 714/2009 of the European Parliament and of the Council of 13 July 2009 on conditions for access to the network for cross-border exchanges in electricity, *Official Journal of the European Union,* no. L211/15.

4

Metering: A Core Activity of the DSOs

4.1. Smart meters are key tools for the deployment of smart grids

In most European Member States, the installation of smart meters will be the responsibility of distribution system operators (DSOs). As part of the physical grid infrastructure, the meters fall within the grid operator's domain, *making the DSOs best suited to manage the smart metering infrastructure.*

The rise of smart metering systems in Europe today is fostered by the European Union (EU) legislation. This includes the Third Energy Package and other legislative instruments such as the Energy Services Directive, the Energy Performance of Buildings Directive and the Energy Efficiency Directive. In detailed provisions on smart metering systems, it is required that end customers are provided with individual meters that are able to accurately reflect consumption and provide information of the actual time of use. Furthermore, they emphasize the adoption of smart meters as a tool both to enhance competition on retail markets and to foster energy efficiency.

Moreover, as already indicated in previous chapters, smart meters greatly contribute to the optimization of the network management. They help for:

– better fault identification and localization on medium-voltage (MV) and low-voltage (LV) networks, ensuring faster interventions and reduced outage duration;

– detailed monitoring of the power quality, thus reducing the number of customers complaints and providing faster answers;

– increased capacity to act remotely on the networks, in particular to manage peak shaving programs;

– implementation of new tools to forecast constraints on the networks;

– better observation and control capabilities.

4.2. A continuous improvement and innovation approach

4.2.1. *From manual to remote reading for mass market customers*

At the beginning of the 1980s, the power line carrier (PLC) technology, attractive and full of promise, was unfortunately not mature enough to be used to a large scale at industrial levels (see section 4.5).

In France, for example, it was decided to dedicate a transmission medium based on PLC communication to the mass market (customers connected to LV < 36 kVA). At short distances, this medium was able to provide access to meter information, at a point available in a permanently public domain: remote reading. In contrast to remote metering, remote reading still requires the readers to move.

The development of remote reading facilitates increasing the reliability of the reading and significantly increases productivity.

A time period of several years was necessary to design, develop, refine and begin deploying electronic meters and associated mobile data acquisition terminals. Hence, at the beginning of the 1990s, the electronic meter was systematically installed in new constructions. After several years of electronic meter deployment, electromechanical mono-phased meters were no longer supplied.

In 2001, Electricité de France (EDF) decided to pursue experimentation with remote metering through PLC, initiated during previous projects, in order to obtain the required knowledge about the operation of an automated meter infrastructure-type (AMI-type) system.

4.2.2. *20 years of smart metering and remote reading for industrial clients*

Sophisticated rates could be established before the transition to electronics, with electromechanical metering systems associated with clocks and 175 Hz relays for the potential periods of moving peaks.

The transition to electronics made possible a technological leap for industrial customers (connected to MV or LV > 36 kVA). Several types of electronic meters were deployed 20 years ago, enabling:

– remote reading of the meters, first by switched telephone network (STN) and then by GSM data;

– measuring, relative to the load prescribed to the seasonal time-of-day segments, the “root mean square value” of potential load excesses in this segment (while before, the occasional power excesses were measured at absolute levels and in periods cumulated over the month), which made it possible to set up more precise pricing for these excesses;

– recovering load curves at 10 min steps.

Experiments conducted in 2001 on the PLC made possible the adaptation of the technology for industrial customers with a balanced business model. Thus, since 2006, remote reading through PLC has been used industrially for professional or industrial customers of the LV network (LV > 36 kVA), when it becomes profitable relative to reading in the field.

4.3. AMI metering systems

Both electromechanical and electronic meters played a key role in power distribution by systematizing the metering of consumed power and by contributing to the invoicing process. Nevertheless, they present several disadvantages, specifically:

– consumption information gathering needs to be done at the source, which is considerably less frequent than necessary for the billing process. Therefore, supplier invoices have to be based on an estimated index, which is the source of numerous customer complaints;

– the metered information is a consumption index and not a load curve; it is impossible for the customers, the DSO and the suppliers to obtain real consumption profiles;

– interventions on meters are generally required during relocation or modifications of the supplier contract; customers need to be present at their home;

– development of dynamic pricing or demand side management (DSM) services is not possible with conventional meters.

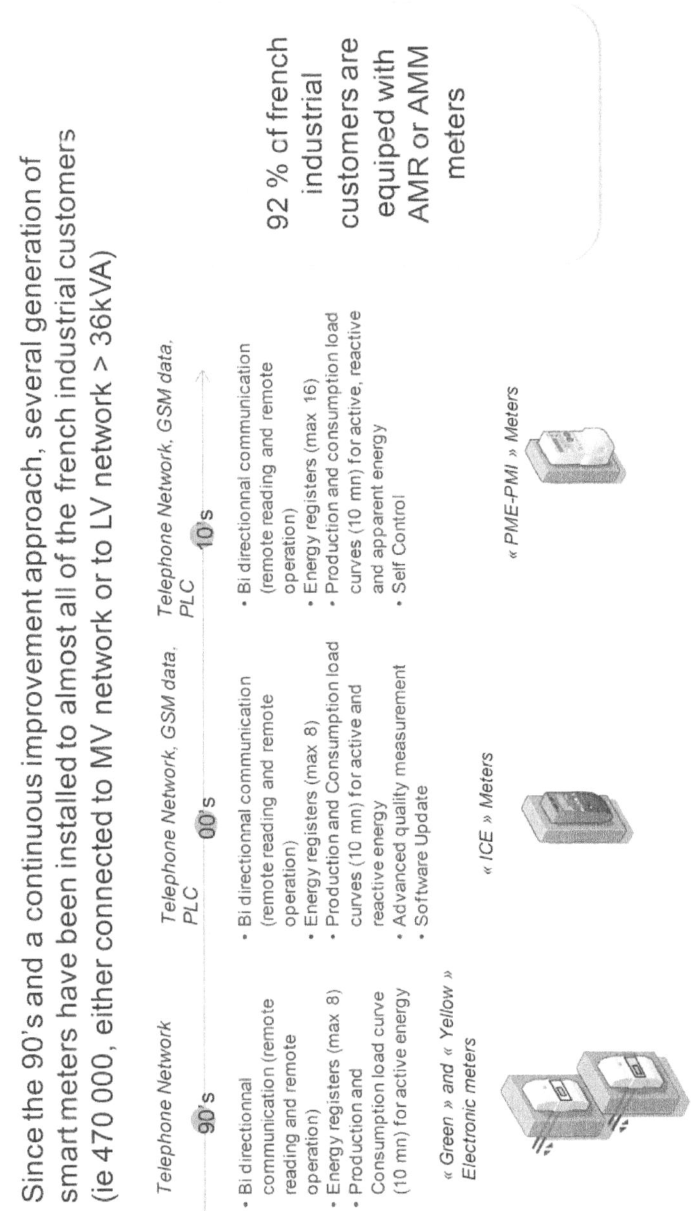

Figure 4.1. *Evolution of the smart meter selection in France for SMEs / SMIs (source: author)*

Since the beginning of the 2000s, under the impulse of public authorities (regulators, department of energy, environmental agency, etc.) and with the aid of technological evolutions, new industrial metering infrastructures have been developed. First, automated meter reading (AMR) systems, based only on unilateral upstream communication, simply collected consumption indices remotely, without bringing a complete solution to metering challenges. Then, systems known as automated meter management (AMM), based on bidirectional communication, were able to collect information and facilitate remote actions on the meter.

AMM systems quickly became the standard, as they provided a response to the overall goals: addressing changes in regulations (variable depending on geographical area and country), improving client satisfaction, contributing to the evolutions expected by the regulator/suppliers, enabling DSM actions and energy transition, improving distribution performance, etc. In this book, we retain the term AMI.

The various international experiences do not establish a single technological standard, whether in the context of meter functionalities (consumption-only metering, prepayment, tariff management, quality measurement, downstream operation, etc.), concentrators (simple hub, measuring element, telecommunications optimization, no concentrator at all, etc.) or central system functions (integrated, specific, level of data processing, interfaces with other IT of the DSO, etc.).

By contrast, two approaches stand out in terms of infrastructure: either an integrated approach from end-to-end or a modular approach based on interoperability.

The integrated approach requires a single manufacturer for both concentrators and meters. Certain equipment manufacturers also suggest software solutions for a "head-end system" (data collection and validation system for the meter). This approach is generally used by small DSOs (several hundred thousand customers), as project operation is simplified by the limited number of interveners. However, once the system is in place and the meter fleet is installed, the DSO is bound in the long term to the manufacturer, without necessarily having control over changes and infrastructure maintenance costs.

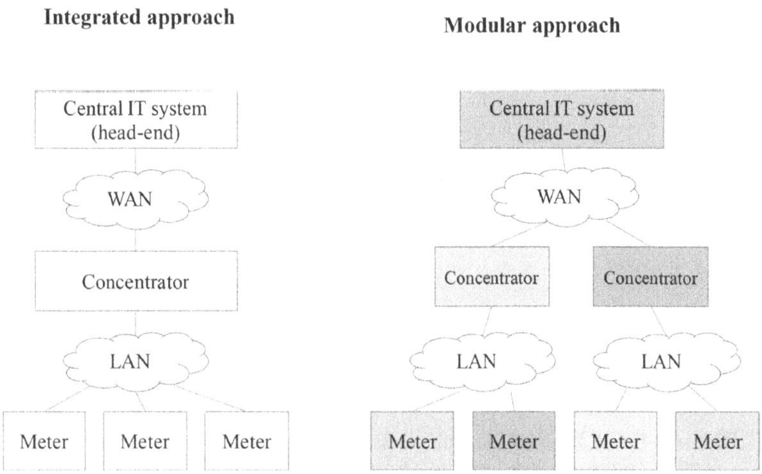

Figure 4.2. *Approach to AMM system infrastructure (source: author)*

The modular approach depends on several equipment manufacturers (meters and/or concentrators) and on an editor/integrator of the software solution providing the "head-end" system. This approach is generally used by large DSOs (several million clients), as it reduces the dependency on any single manufacturer and enables bringing competition in the context of public tenders. However, this approach requires a considerable implication of the DSO in the establishment of technological choices and a strong investment in project operation (design, experimentation and deployment), given the significant number of stakeholders/parties involved.

Communication is at the heart of metering systems (see Figure 4.2), ensuring communication between the meter and the concentrator (local area network – LAN) on the one side, and between the concentrator and the central system (wide area network – WAN) on the other side. In certain configurations, the meter could communicate directly with the central system.

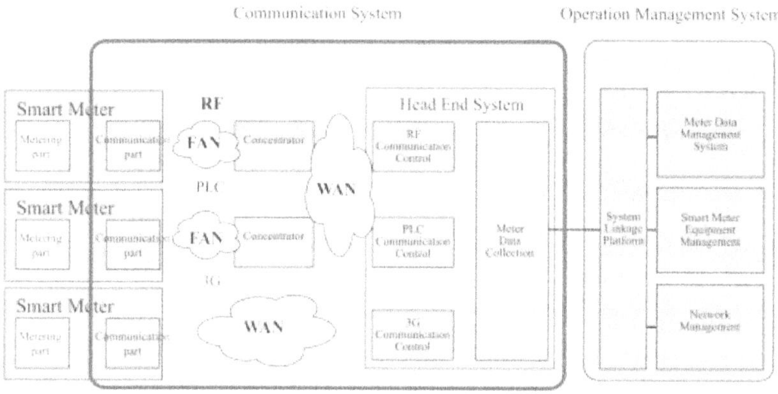

Figure 4.3. *AMI and communication technologies (source: Toshiba http://www.toshiba.co.jp/about/press/2013_05/pr0103.htm)*

As far as LAN is concerned, the DSOs generally opt for one of the three most mature technologies (wireless mesh, GSM/GPRS/3G data and PLC):

– *GPRS is directly used by every meter up to the central IT*: this point-to-point technology, favored by telecommunication operators, was primarily retained by few countries where the metering responsibility lies with energy suppliers (e.g. UK) and where deployment is aimed at "high-value" customers. In addition, it is generally used in high-population density areas. Its principal downside is economical: *the GPRS solution turns out to be twice as expensive in investment and operation as the CPL.* Furthermore, another disadvantage resides in the quick obsolescence of these technologies, which leads to a shorter lifespan for the modems; outside the cost of their renewal, the customer impact in terms of image will be very high. Finally, this solution constrains the DSO to an infrastructure that it does not control.

– *The radio mesh solution presents the same disadvantages. It is adapted to lower density areas* (e.g. United States). Although it is cheaper than the GPRS solution, it still has a higher cost than the CPL approach.

– *PLC between meters and concentrators over LV networks: PLC is the perfectly adapted technology if the DSO is responsible for the metering activity, and if bulk and systematic deployment is considered.* On an aside, this is the reason why it was retained in countries where smart metering projects are handled by the DSOs (Italy, Spain, Holland, etc., see Figure 4.3). By relying on the infrastructure that is already financed, *the PLC technology combines the advantages of control over the cost and the infrastructure, with the induced benefits in terms of quality* (absence of dead zones) and security (independence from telecommunication operators and stability of the chosen approach).

In the context of the WAN, the DSOs primarily opt for GSM/GPRS/3G data, as the telecom operators' infrastructures generally offer a good rate of coverage over national territories. In certain countries, generalized deployment of optical fiber cable could create an alternative (e.g. the goal of the digital European agenda is to enable 50% of the population to have access to high-speed Internet (100 Mbps) by 2020).

4.4. Focus on Linky smart metering system

4.4.1. *Scope of the project*

The Linky project scope is in strict adherence to the architecture of the system. *It is, therefore, limited by the components and communication technologies located between the meter and the central information system (CIS) and hence integrates neither the downstream scope of the meter, nor the other information systems (ISs) of the DSO, which the CIS already works with or will have to work with.*

	Deployment	Generation	Software download	Downstream communication functionality	LAN technology used	Guaranteed interoperability	Breaker
Architecture	80% in 2020	AMM	★★★	★★★	Country dependent	★★★	★★★
ERDF (France)	1%	AMM	★★★	★★★	CPL	★★★	★★★
E-On (Sweden)	100%	AMR then AMM	★	★★	1/3 GPRS 2/3 CPL	–	★
ENEL[1] (Italy)	100%	AMR then AMM	★	★★	CPL	–	★
ENDESA (Spain)	25%	AMM	★★★	★★★	CPL	–	★★★
IBERDROLA (Spain)	10%	AMM	★★★	★	CPL	★★★	★★★
ENEXIS (Holland)	15%	AMM	★★★	★★★	GPRS and CPL	*N/A*	*N/A*

Table 4.1. *AMI architectures in primary European countries (source: Berginsight 2011)*

1 *Source*: Enel, London presentation, January 2013. All functionality evolves toward the state of the art with the transition of countries into AMM.

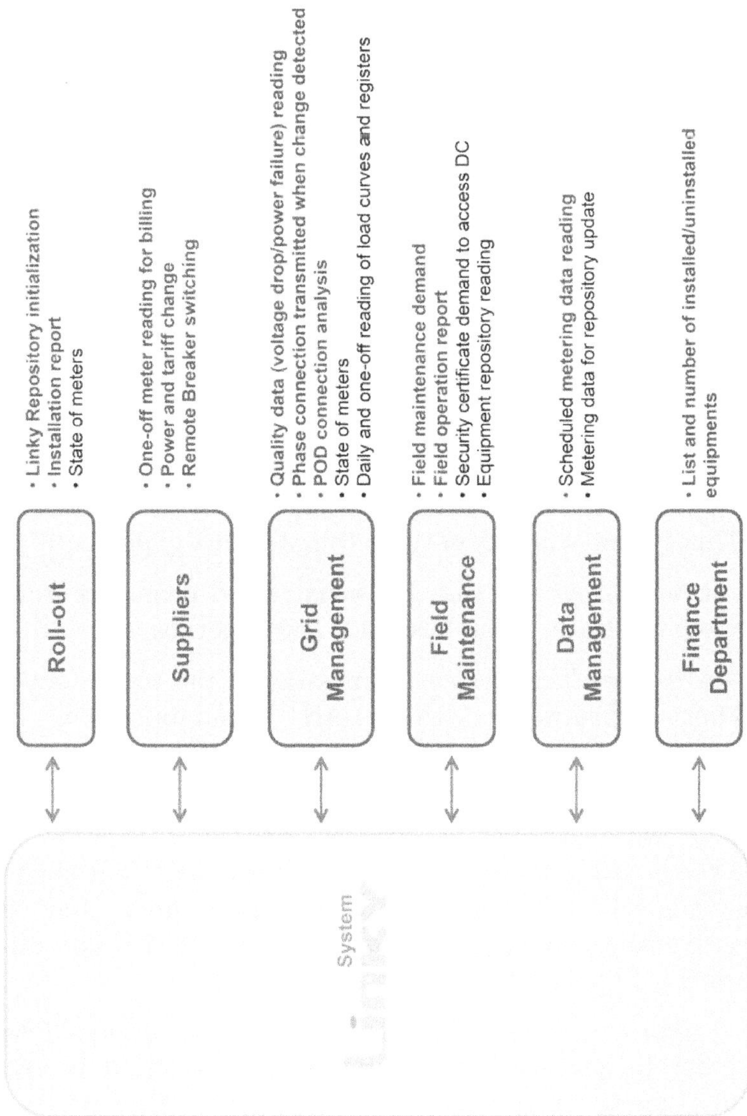

Figure 4.4. *Main types of service: data providing / reading, remote operations on meters*

Hence, outside of the Linky project scope, the projects of renewing the mass market customers' IT and the system for storing consumption data are developed in parallel with their own steering group (planning, costs, risks, etc.). Furthermore, issues downstream of the meter are not part of the project scope. For instance the Linky meter's specifications were changed[2] to take into account the requests coordinated with the parties involved, but *the Linky radio emitter (LRE) is not within the Linky project scope*.

4.4.2. *Architecture and technical choices*

4.4.2.1. *The Linky system: from the central information system to the meter*

This system consists of three complimentary parts:

– *a central system*, the CIS, constructed primarily from a services interface, an orchestrator and supervision;

– *a decentralized system*, consisting of the meter and its software (firmware), the LAN communication, the concentrator and its software (firmware and a unique software package (USP)). The most structuring choice is related to USP. This enables interoperability functionality;

– *the network operation center (NOC)*: a node that ensures the connection between the central system and the decentralized system by using GPRS. There are two main reasons behind choosing to use an NOC:

- to provide a multioperator telecommunication solution in order to guarantee better access to the signal in every point of the territory;

- to constantly keep the benefit of competition between different telecom operators.

2 Removal of the USB port on the front of the meter in favor of a normalized port under the hood of the meter, in an unsealed area accessible to the client, to accommodate a radio emitter.

4.4.2.2. *Choosing PLC for LAN*

In the context of communication technologies, as confirmed by benchmarks run on the AMI projects, three generic solutions were possible (radio mesh, GPRS or PLC).

ERDF's choice was, in terms of PLC standards, the CENELEC A spectrum (5–95 kHz), for four reasons:

– lower costs of electronic components compared to higher frequencies;

– lower modem consumption (therefore, also lesser cost);

– less modem heating (therefore, simpler meter design in small spaces);

– guaranteed availability of the spectrum, reserved for the DSOs.

This was to provide identical carrying for all services (remote collection, remote interventions, etc.).

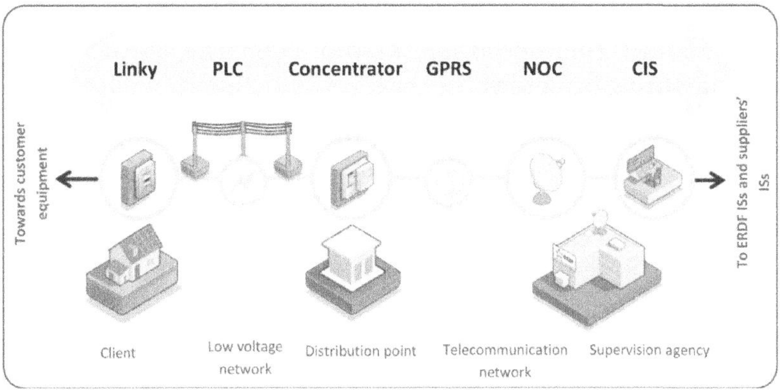

Figure 4.5. *Architecture of Linky system (source: ERDF – presentation "The Linky Project in France" at Africa Smart Grid Forum 2014) http://www.africasmartgridforum2014.org/fr/expert/sessionb2/marc-delandre-projet-de-compteur-intelligent-lynky-en-france-fr.pdf*

4.4.2.3. *Deployment plan*

Generalized deployment is planned with 3 million meters to be installed between 2015 and 2016. Over this period, the majority of the meters will use the G1 PLC protocol, which has proved itself during experimentation with 300,000 meters deployed in Lyon and the Tours region.

The protocol and its equivalents today constitute the primary PLC reference worldwide. Indeed, the closest protocol is Meters and More, which is the foundation for the 30 million AMR meters deployed by Enel and for the program of 13 million Prime meters which are being deployed by Endesa (approximately 3 million meters deployed by the end of 2013).

ERDF will also use the new G3-PLC protocol, whose generalization will take place for this occasion in four stages:

– test G3-PLC in the field;

– ensure interoperability between G3 meters and G3 concentrators;

– validate the operation of G3 over the entire Linky System with G3 meters;

– secure the bulk deployment process of G3 equipment.

The goal is to pursue deployment up to 35 million meters featuring G3 technology between now and 2021.

4.4.3. *A point on system operation*

4.4.3.1. *Primary elements of the Linky meter*

The display facilitates the consultation of the index, current load and maximal load reached from the beginning of the day.

The meter manages two tariff calendars: the supplier tariff calendar (with 10 possible indices) and the DSO calendar (with four possible indices). The meter provides one physical dry contact and seven virtual contacts driven by the supplier's schedule and transmitted by the tele-information client (TIC).

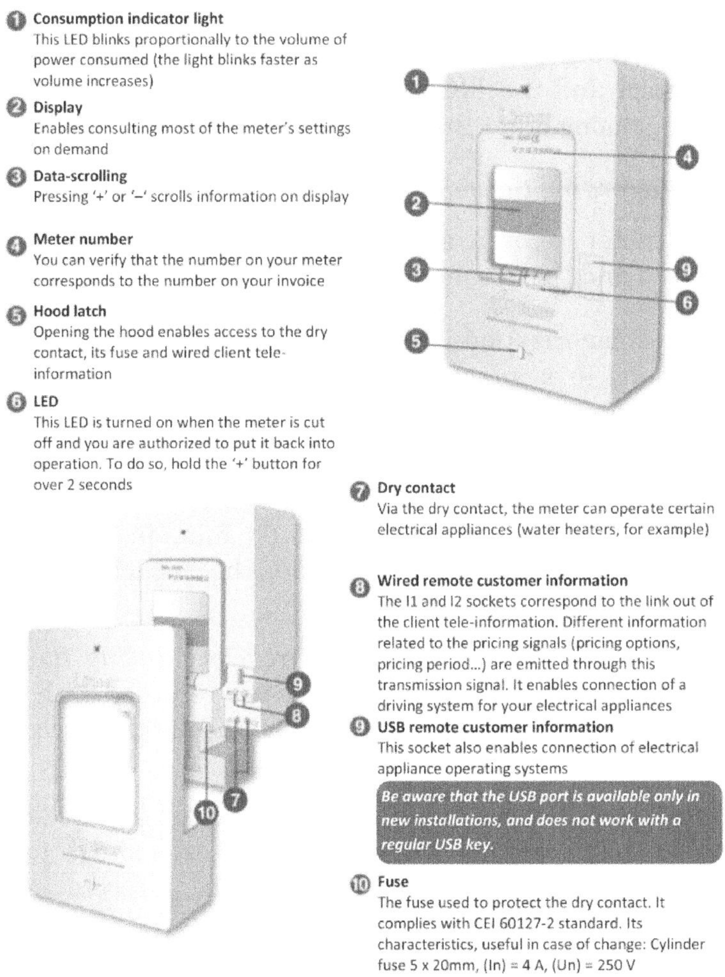

Figure 4.6. *Functionalities of Linky meter (source: ERDF, http://www.erdf.fr/medias/Linky/ERDF_Linky_Notice_Compteur_Monophase_1.pdf)*

The meter manages two distinct notions of power load: the load prescribed by the customer's contract and the maximum power load. In general, the maximum load is equal to the subscribed load. The meter constantly measures power consumption. Load control is assured by checking the maximum power load and following a mode of operation similar to circuit breakers.

Downstream operation of the meter is possible with the dry contact and the TIC (with transmission of virtual relay position, among other things).

4.4.3.2. *Recording the values*

The meter records every day, at midnight, the values measured and these values can be collected by the system.

The memorization capacity of the Linky meter corresponds to a given amount of saved points. Once this number is reached, the oldest events are erased in favor of the latest events. In particular, this meter records:

– *the load curve*: Linky enables recording of the load charge at a step between 10 and 60 min, in withdrawal and/or injection; in consumer mode, the capacity is 3,600 records for the withdrawal load curve;

– *the apparent maximum power reached during the day*: the apparent daily power maximums are recorded with their withdrawal and injection time stamps; the depth of record is 63 daily saves;

– *the prolonged and brief outage monitoring*: a dedicated journal that records all events corresponding to an outage (outage and recovery); the depth of record is 100 events;

– *voltage excursions*: 150 records enable voltage excursions to be monitored (date and duration of the voltage excursions outside of the regulatory range);

– *the movements of the switching device*: a journal dedicated to recording circuit opening and closing, the reason and the switching device. Its depth is 20 records. On a side-note, for every action on the switching device, correctness of operation is controlled;

– *a device for detecting anomalies and autodiagnostic functions*: the meter integrates anomaly detection and self-diagnosing devices. These indicators are recorded, in case of a problem, in an error registry. This status registry also gives the state of certain meter functions in real time.

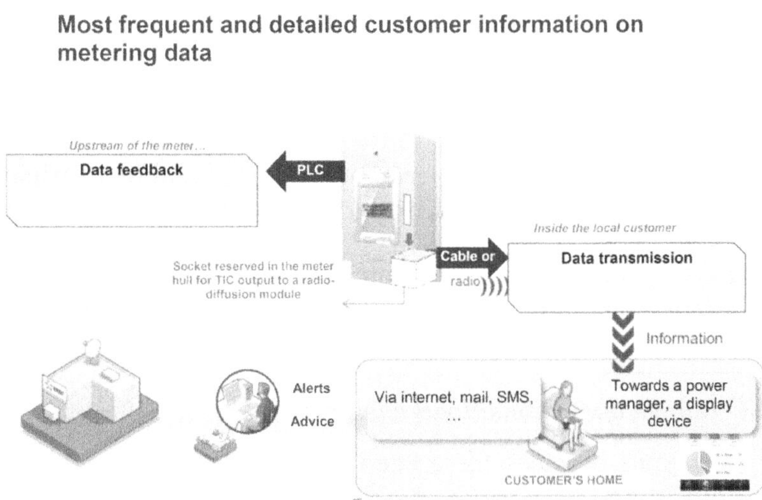

Figure 4.7. *Upstream and downstream communication modes of Linky meter (source: ERDF – presentation "The Linky Project in France" at Africa Smart Grid Forum 2014)*

4.4.3.3. *System functionality*

Systematic or on-demand collection

Record collection: all values recorded by the meter can be collected on-demand.

The indices: the indices are collected systematically every day.

Load curves: load curves are recorded at 10, 30 or 60 min steps and collected daily (within the capacity limitations of the system), after the subscription request.

Collection during events

Anomaly detection and self-diagnosis device: the collection is made automatically in case of voltage excursions, outages or anomaly detection.

Remote supplier operations

Operation and adjustment of the switching device inside the meter: remote-controlled switch; switching device with a remotely configurable trigger threshold.

Pricing parameters: remotely modifiable pricing parameters (subscribed power and tariff schedule).

Mobile peak rates: the system provides communication of orders for mobile peak transition for customers with contracts with mobile peaks rates.

DSO remote operations (load shedding)

Two types of cases have been identified: complete facility load-shedding (anticipated outage) and load-shedding through output reduction.

The command corresponding to this function defines the dates and hours of the beginning and the end of a load-shedding interval.

For load-shedding through output reduction, this command also indicates the percentage decrease that needs to be applied to the breaking capacity.

4.4.4. *Scalability and security of the Linky system*

All system versions underwent performance tests based on, first, a projection method and then on simulation servers equivalent to those currently used in generation. Two parts of the system need to be scalable: the CIS, to process data collection, teleoperators, flow exchanges with other. ITs (customers, metering) and the NOC, to guarantee the absorption of the information flow from the CIS toward the concentrators and *vice versa*.

The security of the Linky system is based on two main principles. The first principle concerns the authentication, confidentiality and integrity of exchanges. It relies on an asymmetrical key encryption technology between the concentrator and the CIS (usage of certificates) and a symmetrical key encryption technology between the meter and the concentrator. The second principle is the securing user tools (supervision and administration teams).

In particular, the following security requirements are anticipated:

– prevention of malicious simultaneous opening (or closing) of a large number of switches in the meters;

– protection of the private consumer life and prevention of unauthorized access to commercially sensitive information (CSI);

– guarantee of continuity and quality of ERDF/metering activity;

– protection of access to IT.

General information security of the system was specified and is the state of the art:

– every specified function was tested and works adequately, including LAN security (security between meter and concentrator);

– security audits of the “white box” (the auditor has the specifications and certain passwords) and “black box” (the auditor does not have the specifications or any passwords) type were carried out (PLC and meter are outside of scope).

– WAN security was validated with telecommunication operators.

Taking security into account relies on the principle of permanently adapting to the state of the art.

4.4.5. *Techno-economic analysis*

Analysis was carried out to determine the significance of the Linky project to the community and its impact on customers’ invoices.

The economic analysis shows that the Linky project is economically balanced to the scale of the distribution activity.

This economic balance ensures, in the long term, the neutrality of the project on the customer invoice. The regulatory system in place minimizes the impact of the project on the first TURPE (*Tarif d’Utilisation des Réseaux Publiques d’Electricité* – Public Power Network Operation Tariff) theme period, in order to make it concurrent with the network gains, without taking into account the additional potential gains that the consumers may be able to benefit from beyond the distribution scale threshold (up/downstream optimization, DSM, etc.).

This balance is based on the techno-economic analysis that compares the “business as usual” scenario (without Linky) with a scenario where Linky is deployed in order to estimate the significance of this project. The analysis leads to an economically balanced net present value (NPV). This NPV is justified by the fact that the overhead of the investments and the exploitation costs is compensated by gains on non-technical losses, small interventions (virtualization of an important part of these), data collection and network gains (improvement of operation, phase balancing, investment optimization, etc.).

4.5. Focus on G3-PLC technology

4.5.1. *Communication principles of the power line carrier*

PLC communication uses a higher frequency, low-energy signal, which is superimposed on the alternative electrical current at 50 Hz in the power network. It can be transmitted through MV and LV networks, and downstream of the meter.

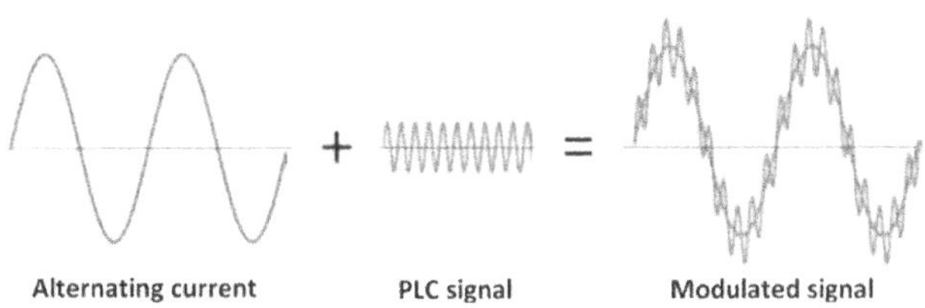

Figure 4.8. *Depiction of a signal modulated by PL communication (source: Deliberation of the Commission of energy regulation on 12 July 2014, about recommendations on development of smart low-voltage electricity network)*

The PLC signal enables transmission of information over the entire electrical network. It is received and decoded by equipment featuring a PLC receptor of the same category as the emitter, located on the same network.

4.5.2. *Different types of physical level PLC modulation technique*

The different technologies of PLC are defined according to the modulation techniques. Digital modulation could use one or more of the three parameters of a sinusoidal signal:

$$s(t) = A \cdot \cos(2\pi \cdot fc \cdot t + \theta k)$$

– ASK: amplitude-shift keying;

– PSK: phase-shift keying;

– FSK: frequency-shift keying.

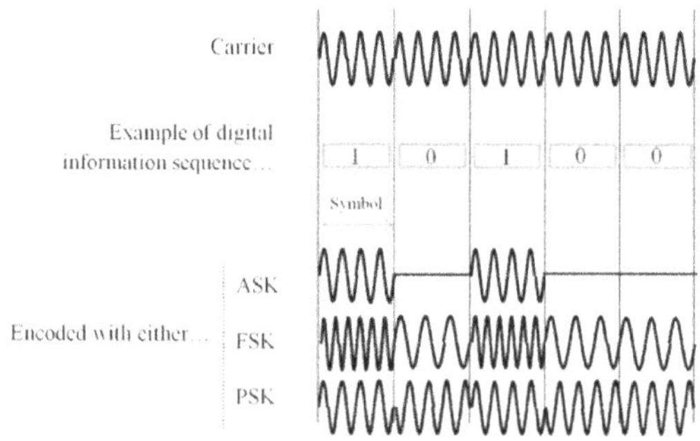

Figure 4.9. *Representation of different PLC modulations (source: author)*

FSK operating mode

The emitter affects the frequency Fs at the value 0 and the frequency Fm at the value 1.

The Fs and Fm frequencies are separated from each other (spread). By placing the Fs signal far from the Fm signal, the quality of their respective transmissions becomes independent of the narrowband noise, which is frequently found in the network.

The receiver performs a classical FSK demodulation on the two possible frequencies (the half-channels), which generates two demodulated signals. If the average reception quality (signal-to-noise ratio (SNR)) of the two half-channels is similar (see Figure 4.11), then the decision unit opts for the higher demodulated channel.

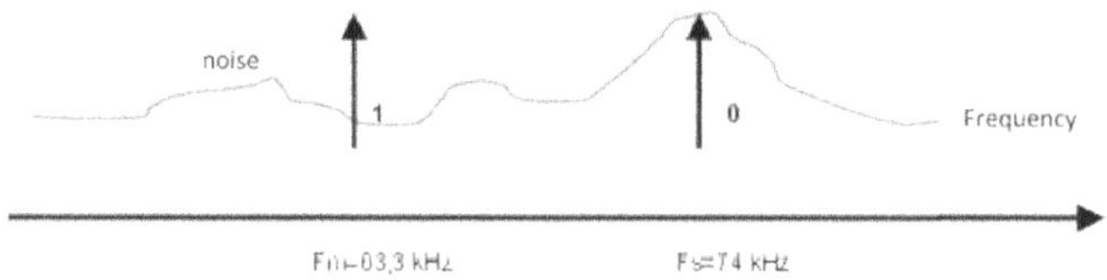

Figure 4.10. *FSK operating mode*

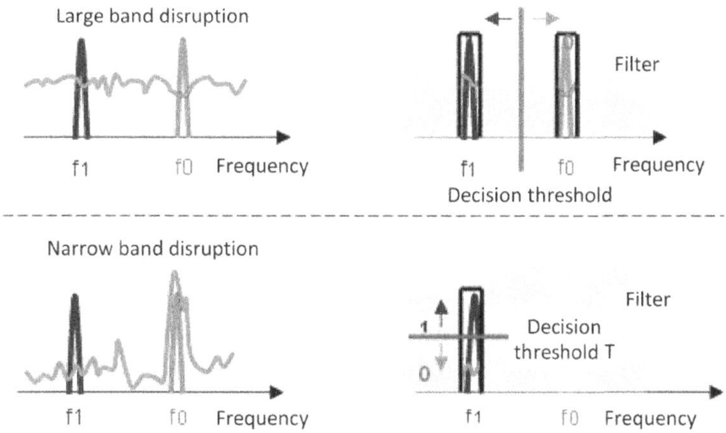

Figure 4.11. *Disruptions management in FSK mode (for a color version of the figure, see www.iste.co.uk / boillot / smartgrids.zip)*

ASK operating mode

If the quality of average reception of one of the half-channels is better than the other, the decision unit compares the demodulated signal of the better channel to a threshold T, ignoring the other channel.

PSK operating mode

Data bits are mapped for differential modulation such as differential binary PSK (DBPSK), differential quadrature DPSK (DQPSK) and D8PSK.

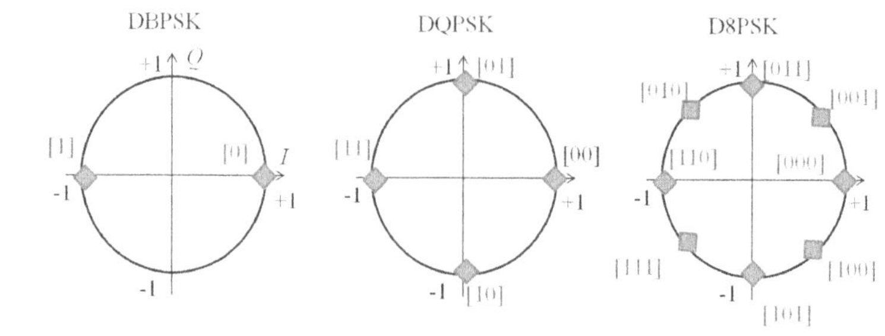

Figure 4.12. *PSK working modes (source: author analysis)*

Spread frequency-shift keying (S-FSK) uses a combination of FSK and ASK to reinforce the performance of FSK.

Orthogonal frequency division multiplexing (OFDM) uses a combination of ASK and PSK. The available bandwidth is divided into a number of subchannels, which can be viewed as many independent PSK modulated subcarriers with different non-interfering (orthogonal) subcarrier frequencies.

	Low data rate	Medium data rate	High data rate
Data rate	0–10 kbps	10 kbps–1 Mbps	>1 Mbps
1.1.1	1.1.2	1.1.3	1.1.4
Modulation	BPSK, FSK, SFSK, QAM	PSK+OFDM	PSK+OFDM
Standards	IEC 61334, ANSI/EIA 709/1.2, IPB	PRIME, G3, P1901.2	G.hn, IEEE1901
Frequency range	Up to 500 kHz	Up to 500 kHz	In Mhz
Applications	Control and command	Control and command, voice	Broadband over power line, home networking

Table 4.2. *PLC technology classification on the basis of data rate (source: author analysis)*

4.5.3. *The characteristics of G3-PLC technology*

The power line channel is very hostile for PLC communication. Channel characteristics and parameters vary with frequency, location, time and the type of equipment connected to the power network.

The lower frequency regions from 10–200 kHz are especially susceptible to interference. Besides background noise, it is subjected to impulse noise and narrowband interference and channel impulse repose time of up to several hundred microseconds.

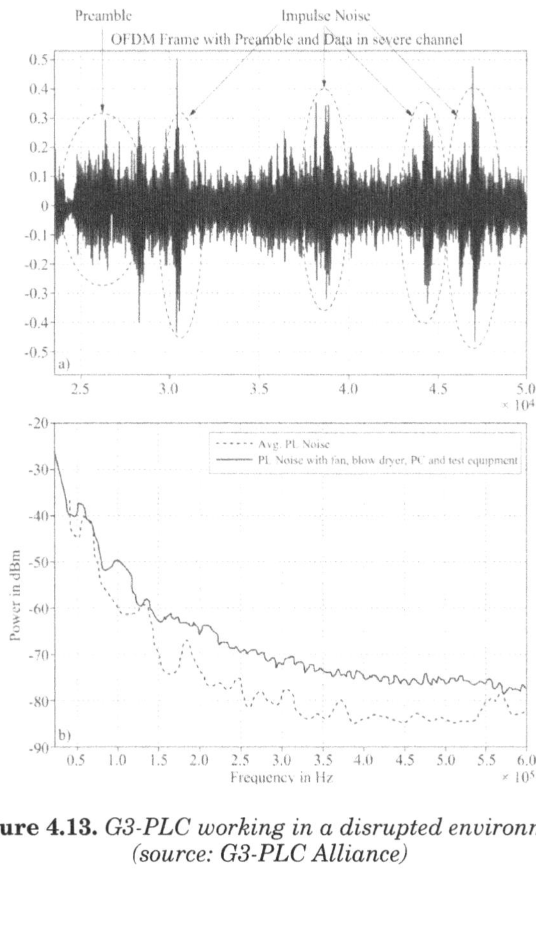

Figure 4.13. *G3-PLC working in a disrupted environment (source: G3-PLC Alliance)*

G3-PLC provides effective, efficient and secure communication over power lines. The G3-PLC technology is composed of five layers:

– the physical (PHY) layer which supports an internationally accepted band from 10 to 490 kHz (FCC, CENELEC and ARIB);

– on top of the PHY layer, the G3-PLC MAC and 6LoWPAN layers provide the conditions for fast and secure communication by means of advanced routing techniques through hopping *via* devices connected to the network;

– on top of the MAC layer, transport and network layers enable IPv6 and potential services (simple network management protocol (SNMP), trivial file transfer protocol (TFPT), etc.).

4.5.3.1. *PHY layer is based on OFDM*

OFDM is a method for encoding digital data on multiple carrier frequencies. As explained in the previous section, it divides the available bandwidth into subchannels, which can be viewed as many independent PSK-modulated subcarriers with different non-interfering (orthogonal) subcarrier frequencies. In the CENELEC A BAND, there are 36 subcarriers available for communication.

In OFDM modulation, those frequency bands are chosen for communications that have a high enough SNR to enable the reliable transmission of signals.

G3-PLC supports different OFDM modulations (DBPSK, DQPSK, D8PSK, etc.). The primary advantage of OFDM over single-carrier schemes is its ability to cope with severe channel conditions without complex equalization filters, such as attenuation of high frequencies in a long copper wire, narrowband interference and frequency selective fading due to multipath.

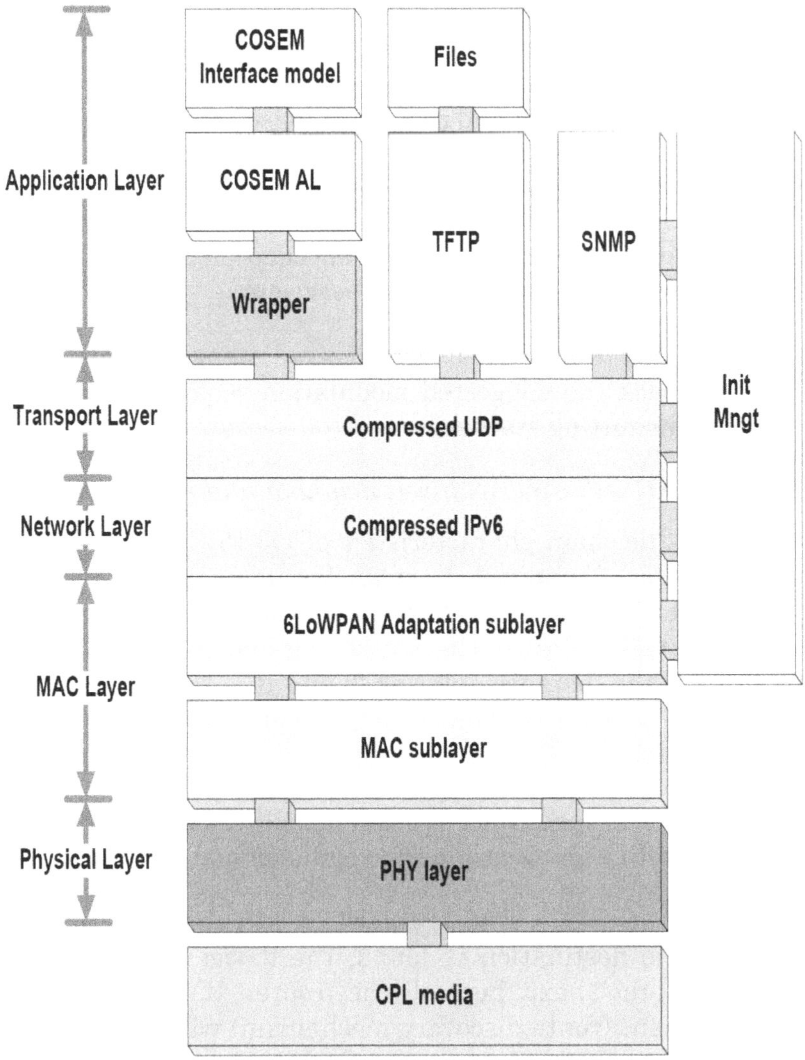

Figure 4.14. *Complete PLC modem (from the PHY to the application layer) (source: G3PLC Alliance)*

G3-PLC PHY uses multilayer error encoding/decoding with convolutional and Reed Solomon coding that provide redundancy bits allowing the receiver to recover lost bits caused by background and impulsive noise.

The OFDM signal is generated by performing Inverse Fast Fourier Transformation (IFFT) on the complex-valued signals produced by differentially encoded phase modulation that are allocated to individual subcarriers.

Based on the quality of the received signal, the receiver will feed back the suggested modulation scheme to be used by the transmitting station.

4.5.3.2. *MAC + 6LoWPAN layer: neighbor and routing tables*

One of the main characteristics of G3-PLC technology is that it uses "hopping" to transfer messages from the "transmitter" (T) to the "receiver" (R). This is necessary when the G3-PLC signal is not strong enough to reach the receiver: if the distance between T and R is too large and/or if the quality of the transmission channel between T and R is not good enough.

Thus, the "transmitter" T uses one additional hop (e.g. via another G3-PLC device connected to the distribution grid).

When a device receives a packet, it will check its routing table. If the destination is found, the device will send the packet to the next hop of the route. If not, a route establishment (route discovery mechanism) will be triggered to find the route to destination. Each node in the network (devices equipped with a G3-PLC modem) has a neighbor table and a routing table.

The neighbor table consists of the fields listing in Table 4.3.

Short address	The MAC short address of the neighbor node
Payload modulation	Differential or coherent mode (coherent mode supported in the FCC and CENELEC A bands)
Tone map	Parameter that defines which frequency subband can be used for communication with the device
Modulation	Defines the modulation type to use for communicating

Table 4.3. *Different routing domains*

4.5.4. *G3-PLC is a mature standard*

4.5.4.1. *Specifications: main achievements*

– The "PHY+MAC" part of the G3-PLC specification is mature, and interoperability tests have been performed for the three market bands (CENELEC A, ARIB and FCC) based on the fixed 2012 version.

– The G3-PLC specification has been published in several "backward compatible" steps and amendments.

– A new stable version has been accepted for December 2013 ITU publication: ITU-T G.9903 (2014).

– It also includes the latest correction and clarification items taking into account the tests performed in June and September 2013, as well as some backward compatible enhancements allowing an extreme degree of convergence for the FCC band between the G3-PLC Alliance reference version, ITU G9903 and IEEE P1901.2.

4.5.4.2. *Interoperability tests and plug fests*

The mission is:

– to develop the process and tests for interoperability: the target is PHY, MAC and 6LowPAN including routing;

– to execute the first rounds of interoperability;

– to validate the specification, to propose correction and clarification of the specification when needed;

– when completely defined, to transfer the process and the tests to certification labs to be chosen by the G3-PLC Alliance;

– to maintain the process and update the tests according to the evolution of the specification.

4.5.4.3. *Market overview*

The G3-PLC Alliance continues to attract members, and the market potential is growing. There is still a tremendous opportunity for a technology that is offering a balance of high data rate, robustness, security, quality of service and low infrastructure cost:

– Europe is moving on, and utilities are making decisions to fulfill the requirements of the European Directive for 2020: ENEXIS and ERDF have made announcements for 2 and 32 G3-PLC million meters to be installed before 2020.

– Asia is the fastest growing market for smart grid applications and a communications technology that works well in noisy environments. Successful field trials have proved that G3-PLC is the answer, and announcements have been made in Japan (CEATEC event, October 2013), mainly regarding the Route B: communication between a smart meter and home energy management system (HEMS). This will represent a market of 80 million modems in smart meters, and probably almost the same in the premises of the customers. In addition, many other Asian countries are interested in G3-PLC, and are working on pilots: Taiwan, Korean, Malaysia, India, etc.

– Other areas are promising: Middle East and South America (Brazil) will need attention from the G3-PLC Alliance.

Figure 4.15. *Members of the G3-PLC Alliance (June 2014) (source: G3-PLC Alliance)*

4.6. The contribution of smart meters for the development of advanced smart grids

4.6.1. *France: Linky at the service of the distribution network*

4.6.1.1. *A dedicated program*

From the inception of the smart meter project, a specific project was launched to develop the functionality of the network around five objectives:

– *enrich and make more reliable the customer connection* and detect automatically changes in LV network usage patterns.

In the geographical information system (GIS), every client is connected to a section of the secondary substation. In the AMI topology, every meter is connected to one of the three LV phases of the secondary substation. The process of phase acquisition and detecting deviations between the GIS and the AMI is industrialized (connection, opening points, temporary reparations, crosstalk, etc.).

Approximately, 5% of the clients are not connected to the correct substation. In the simple cases, the correction times are under 3 min. For more complex cases, we can use load curves or search around the usage of voltage measures made by the AMI.

– *Using data issued from Linky* (outages and voltage excursions) to identify low-quality-of-supply-customers (LQS) and optimize the solutions to treat them, improve the handling of customer claims and enable publication of internal and external statistics.

The first aim is to have at hand, during the deployment of the AMI, a tool for following voltage levels and the occurrence of prolonged and brief outages, by making use of the data issued by the meters. The second aim is to provide businesses with the necessary backing, assorted information, organized and hierarchized in the most appropriate IT to improve the handling of claims and the intervention and investment decisions.

– *Specify the new secondary substations*, as well as the policy and the equipment, which enable the evolution of the existing facilities, aerial and underground, in consistency with the Linky project.

– *Reduce outage times* with the remote collection of failure detectors of the substations equipped with Linky concentrators. Improve the detection of LV network failure and move forward the handling of customers by helpdesks.

Optimize the current load shedding device by providing more flexibility.

The TCC has the possibility of checking the meter and obtaining the meter's status and error entries. These entries contain information such as the state of the switch or the cause of the opening of the switch (overvoltage, overcapacity and IS orders). The "Linky ping" is a diagnosis aid and is faster than asking the client to check the meter.

– *Improve investment decisions* by keeping account of the customer connection phase and the actual load curves of the customers.

The new information accessible via AMI (connection phase of the clients and actual load curves of the customers) makes possible to explain atypical situations of incident or supply quality, reconnecting new facilities in an optimized manner (prioritizing the connection phase in order to limit losses) and optimize processing models used in conventional studies (enhanced modeling of the network).

4.6.1.2. *The networked metering system is an important tool for operation and control of MV/LV networks*

Alerts coming from the downstream by the AMI in case of communication failure between the concentrator and one or several meters from a set are possible indicators of a LV supply outage.

The cyclical queries to the AMI system associated with the alarm usage related to incidents (opening of a breaker in case of overvoltage) make possible the upstream supervision of the LV network. The performance of the supervision depends on the reliability of the PLC connection and the detection delay of a communication failure between the concentrators and meters.

In case of an overvoltage in the network, the breakers in the meters are supposed to open without the possibility of closure by the customers in order to protect the equipment. It is necessary to send each meter a command to return to the initial state once the network problem is resolved.

The AMI offers a device of last recourse for load reduction.

Targeted load capping or capacity modulation is most comfortable for the customers than a complete load-shredding at the primary substation. It lies in the ability of the DSO to *reduce in a target manner and in capacity the network load.*

This capacity reduction can first and foremost be considered as a lever of last recourse that can be activated in case of *a congestion on the transmission or distribution network anticipated by the forecast management process or detected by operation* and could not be resolved by the intermediary of other network or market flexibilities.

This safeguard device can be activated by the DSO for its own needs or in response to a request from the transmission system operator (TSO).

4.6.1.3. *A contribution to the enhancement of smart secondary substations*

One of the challenges of smart grids consists of being able to increase the intelligence of secondary substations by adjusting the evolution of local issues.

The solution consists of a standard generalized concentrator, featuring a restricted set of network functions in the majority of substations and supplementary network functions integrated into a distinct box connected to the concentrator in a limited number of substations with high stakes.

These functions enable better follow-up and control of voltage levels along MV and LV feeders (see Figure 4.16).

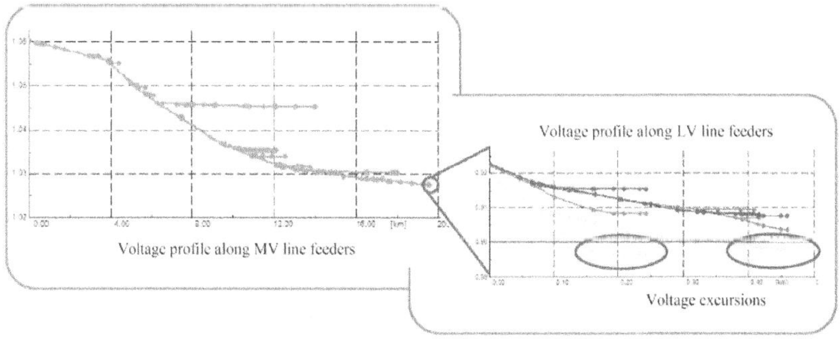

Figure 4.16. *Example of voltage profiles along MV and LV line feeders (for a color version of the figure, see www.iste.co.uk / boillot / smartgrids.zip)*

4.6.1.4. *Developed functions*

Coupling the Linky system with the network operation system and the GIS enables the development of the following functions:

– *Detection of undersupplied customers*: the analysis of voltage excursions recorded by the Linky meters, cross-referenced with the analysis of events that occurred in the network (operations and incidents), facilitates the identification of undersupplied customers (that is clients whose supply quality does not meet regulatory requirements) from real measurements, and not only from calculations and simulations.

– *Following the quality of supply in a given geographical location*: the analysis of the data recorded by Linky meters (outages and voltage excursions), cross-referenced with the analysis of events that occurred in the network (operations and incidents), and the different available measurements, will lead us to make a diagnosis of the quality of the supply to the specified area, and ultimately to better target actions of development and reinforcement of the network.

– *Following the power quality*: the domain of power quality (flickers, rapid voltage variations, harmonics, etc.) constitutes another facet of this function. The analysis of the latter is based on the function of operation monitoring.

– *Modernizing the customer information and improving internal reporting*: the available data, more numerous and more precise, enhance the information for the clients and in general the reporting on supply quality.

4.6.1.5. *The demos in progress (some targeted illustrations of the link between metering and network)*

Several demos of ERDF smart grid projects are based on the deployment of Linky meters and concentrators (see Figure 4.17) to develop advanced functions for observation and control of MV and LV networks.

In more detail:

– *NiceGrid project* (Carros): notably, studies the contributions of Linky to identify capacity constraints in high PV generation density areas and to test battery storage solutions.

– *GreenLys project* (Lyon, Grenoble): notably, studies the contributions of Linky to network operation, for investment planning and for detection of non-technical losses.

– *SO GRID project* (Toulouse): demo which aims to test a complete chain of G3-PLC, from the meter to the primary substation, and to assess its performance for the requirements of the network.

– *Smart Electric Lyon project* (Lyon): aims to show the contributions of new Linky functionalities for driving uses upstream of the meter; in particular, it provides works in standardization and usage tests of a Linky radio emitter.

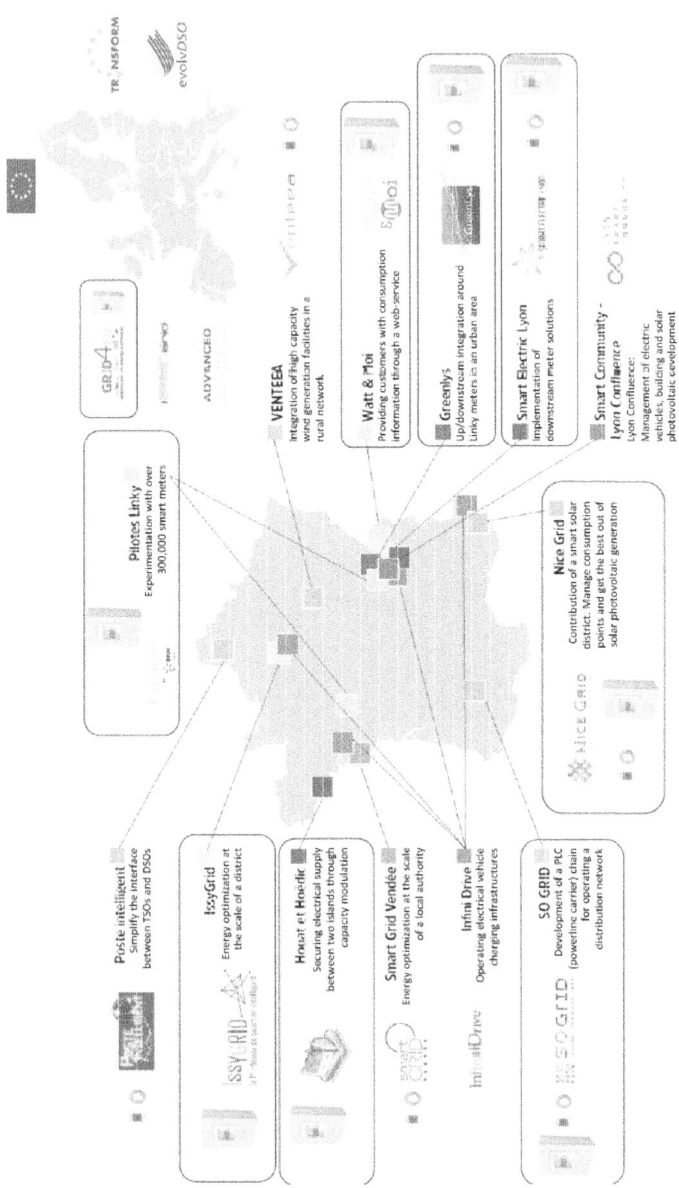

Figure 4.17. *Cartography of ERDF smart grids displays and the link with smart meters (source: ERDF, presentation Smart Grids Paris 2014) (for a color version of the figure, see www.iste.co.uk/boillot/smartgrids.zip)*

– *Houat and Hoëdic project* (Brittany islands): successful experiment in contributing to modulating capacity during maintenance and incidents on the supply cable to the two islands.

Finally, in certain displays, the SME/SMI meter is used either to record the consumption/generation of LV > 36 kVA clients and MV clients, or for the instrumentation needs of the demo – measurements in the secondary substations.

5

Focus on Flexibility Options

5.1. Flexibility, a complementary tool for DSOs

5.1.1. *Introduction*

To limit the increase of network load related to the development of distributed generation, electric vehicles and the "natural" raise of the public distribution network (PDN) peak, the distribution system operator (DSO) needs to implement flexibilities to defer certain network reinforcements.

System operators (transmission system operators (TSOs) and DSOs) have the responsibility to ensure system stability and security of supply. Today, the main tool used by DSOs to overcome increases in electricity consumption, or generation in their network, is to reinforce the grid by laying down more electricity cables, upgrading transformers, etc. But making the most of the flexibility offered by grid users is the approach that is currently gaining momentum.

DSOs could make the most of their network provided that they are allowed to use system flexibility services. In a similar way, increasing flexibility in the electricity market (when technically and economically appropriate) would result in numerous benefits for DSOs, grid users and society as a whole. However, this implies that distribution networks are planned differently, incorporating new risk margins and uncertainty, and are not managed as they used to be but rather are now managed as networks with enhanced observability, controllability and interactions with market stakeholders.

The DSO will procure flexible system services to plan and operate its network in the most cost efficient way. *Flexibility is a complementary tool* through which this can be accomplished. System flexibility services are not a substitute for planning, connection and access investments, but a tool with which to bring more efficiency to the system.

Flexibility, in this respect, can be defined as the "modification of generation injection and/or consumption patterns in reaction to an external signal (price signal or activation) in order to provide a service within the energy system". The parameters used to characterize flexibility include the amount of power modulation, the duration, the rate of change, the response time, the location, etc.

Using system flexibility services for voltage control and congestion management could provide clear benefits for DSOs.

5.1.2. *DSO needs in terms of flexibility*

5.1.2.1. *Frequency control*

The objective of frequency control is to keep frequency within the acceptable range. It is the responsibility of TSOs. With the development of renewable energy sources (RESs) on distribution networks and an evolution toward active distribution networks, active power reserves are needed to meet unplanned increases in demand or sudden losses of production. Today, these reserves are mainly provided by centralized generation connected to transmission networks. However, in some countries with high levels of RES penetration (mainly wind and photovoltaic (PV) energy) some sort of reserve capabilities are now required from RES.

Since RESs are mainly connected to distribution networks, it may become the *responsibility of the DSO to ensure that RESs contribute to active power reserves and frequency control*. In the same way, with the development of active demand (AD) on distribution networks, the contribution of residential customers, and small or medium commercial and industrial customers, to active power reserves, and some sort of frequency control, might be envisaged. In this case, depending on the regulatory context, the DSO might have an important role to play.

5.1.2.2. *Power flow control and network congestion solution*

In distribution networks, flexibilities (AD, generation) might be used to solve congestions on high-voltage (HV), medium-voltage (MV) and low-voltage (LV) networks, through the modification of loads and/or injection, and therefore the modification of power flows on the networks. These needs appear in "real time" and cannot be solved with an investment, contrary to the optimization of network planning. These flexibilities could be used at different time-frames depending on when the network constraints have been detected.

These congestions might be the result of works, failures or extreme conditions (for example, extreme temperature conditions).

5.1.2.3. *Voltage control and reactive power compensation*

As already presented in Chapter 2, the voltage on distribution networks has to be maintained between lower and upper limits. The voltage profile varies along feeders with load and generation, and therefore voltage control has to be performed.

Among the use of flexibilities, correction of the power factor is the easiest and is often already implemented at large or medium customers' premises or generators, either by obligation (technical grid codes and requirements asked at the time of the connection) or by contractual commitments (with possible penalties). However, AD might also be used to contribute to some sort of voltage control at certain points on the network even if the way to implement it does not seem easy. It will require appropriate monitoring in order to avoid possible adverse effects.

5.1.2.4. *Emergency situations*

Measures adopted to maintain system voltage stability can be:

– increase reactive power injection;

– reduce loads;

– start new generation units;

– block transformer tap changers at HV/MV substations.

The DSO is involved in such actions. The use of flexibilities on the distribution network may be considered. However, voltage fluctuations are a very complex phenomenon, and it might be difficult to establish an appropriate strategy. These situations need an automatic action/control from the DSO.

5.1.2.5. *Restoration / blackstart*

Flexibilities may contribute to network restoration after a partial or complete loss of supply on parts of the distribution (MV/LV substation or larger zone) or transmission network or to a blackstart after a system blackout, thus providing service to the DSO and/or TSO.

Limiting consumption or adapting generation will help generation units and substations (primary or secondary substations) to progressively recover load in the network segments and then allow the reconnection of all parts of the network.

5.1.3. *The value of flexibility*

From a DSO point of view[1], using flexibility services could provide a number of quantifiable benefits, both for network operators and grid users. Among them it is possible to point out:

1) *Optimized distribution network capacity investments*

In cases of high generation, or demand, of electricity, parts of the electricity grid can be subject to congestion due to a limited distribution capacity. Network operators traditionally fix this issue by investing in network reinforcements. Using flexibility can help to *defer an investment or could solve congestion when reinforcing the infrastructure would not be possible.* If flexibility avoids an investment, the value of flexibility then equals the sum of capital expenditure (CAPEX) and discounted operational expenditure (OPEX) of the avoided reinforcement. If flexibility services enable the DSO to defer investments, the value of flexibility can be calculated as the avoided return on capital cost over the deferral duration.

2) *Reduced technical losses*

Transporting a kWh from generators to consumers creates network losses (power dissipation in distribution lines and transformers) that are proportional to the length of the electricity route. Flexibility services can help to reduce losses. *Network losses are already given a value, the value of flexibility then corresponds to the amount of electricity that has not been lost.*

---

1 See [EDS 14c].

3) *Congestion limitation to increase distributed energy flows on the network*

By using flexibility services, DSOs could *better control voltage profiles in areas with a high number of variable sources of electricity*. Flexibility can thus directly benefit grid users (e.g. solar panel owners) who would be able to feed-in more energy to the grid. The value here is determined by avoided investments and maintenance costs in voltage control.

4) *Increased distributed generation hosting capacity*

This point is related to the previous point. By helping to keep the network stable, flexibility services could, in some areas, increase the distributed generation hosting capacity of the grid. The value here is also determined by avoided investments and maintenance costs in voltage control.

5.1.4. *Alliander Smart Grids Cost Benefits Analysis (source: Alliander)*

Alliander performed a study with the conclusion that the main changes in the energy landscape with impact on the distribution network are: electric vehicle (EV), heat pumps and solar panels. The company studied three scenarios:

1) "Business as usual" (BAU);

2) "Steady growth" (SG);

3) "Extreme energy transition" (EET).

For each of these scenarios, Alliander looked at it in terms of new peaks in the grid and how to anticipate them. A bottom-up study was performed for all network components and the load that will occur. In all three scenarios, extra peaks are anticipated in the network by 2030. It ranges from +20/40% (BAU), to +75/230% (SG) and up to +300% (EET).

A set of solutions to deal with this is considered, adding capacity to the grid, managing more actively demand/generation and adding storage. From these three main solutions, five specific solutions emerge of which one (storage) is not yet economical (See Figure 5.1) .

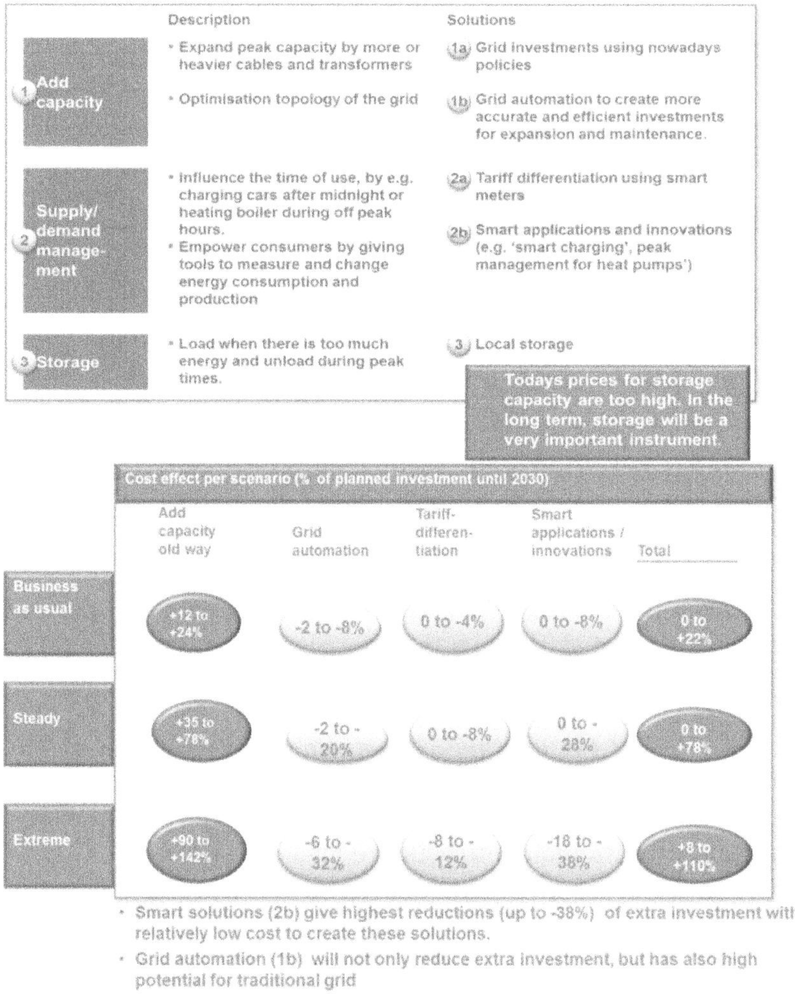

Figure 5.1. *Objectives and conclusions of the study (source: Alliander)*

The conclusion of the study is that using more flexibility instead of only reinforcing the grid is sensible in all scenarios. In the traditional way (EET), this would mean a +142% cumulative investments until 2030. Grid automation, tariff differentiation and smart applications/innovations can reduce this to around +110%.

Smart applications/innovations are smart heat pump, EV smart charging, etc. Tariff differentiation on the supplier side (real-time pricing and time of use) and also on the DSO side (critical peak pricing and critical peak rebate) could reduce peak by an average of 12–20%. The full study results are shown in Figure 5.1.

Smart solutions give the highest reductions of extra investment with relatively low cost. Grid automation has also high potential for the current grid.

5.1.5. *Two major categories of levers can be activated*

5.1.5.1. *Levers related to a punctual and definitive action for optimally positioning RES, for example:*

– *planning the localization of RES*: reduce the part of incorrectly located RES in the distribution network (that is by generating reinforcement constraints);

– *increasing the average photovoltaic power*: decrease the number of facilities connected in LV < 36 kVA, which are replaced by connections in MV and LV > 36 kVA, where connection/reinforcement costs are lower (in €/kVA).

5.1.5.2. *Levers associated with real-time actions, participating in load "balancing" at a local scale, for example:*

– *control of the local peak* (through actions on AD: load shaving, deferment, etc.): limit the peak increase, and therefore the network reinforcement costs.

– *active management of generation*: partially level production (or inversely, stimulate consumption) beyond a certain constraint threshold;

– *reactive management*: injection of reactive power by the generators allows us to stay within the authorized maximal voltage. This injection already takes place in the MV network. This lever suggests that it should be extended to LV networks.

Ultimately, the activation of these flexibilities by the DSO requires analyzing their efficiency and accessibility (see Figure 5.2).

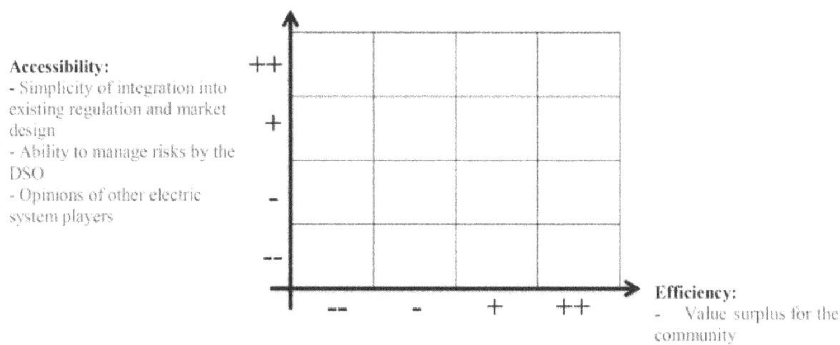

Figure 5.2. *Analysis matrix of DSO flexibilities*

A key point of this analysis is modeling by "*Merit Order*".

5.1.6. *Analysis of the Merit Order*

This modeling is at the heart of quantitative analysis by efficiency criteria. It allows the calculation of the value created or the surplus for the community.

A *Merit Order* is a graph that presents the solutions to a need in decreasing order of economical merit. In abscissa, we find the potential of every energy source (in GW), and their costs (in €/kW per year) in ordinate. By consequence, the

Merit Order is a stepped rising curve (every step being a source).

For a given valorization of system flexibility, the source will remain active while its cost is less than the flexibility value for the DSO. The creation of value for the community is, therefore, equal to the area between the stepped function of the *Merit Order* and the value of system flexibility.

5.1.7. *Information exchange mechanism between DSO and TSO*

According to EDSO, matching system flexibility services and DSO needs will require clear procurement mechanisms. An obvious precondition is for regulation to allow DSOs to use these services. Efficient and secure information exchange will be needed between TSOs and DSOs. Coordination with market players will also be a crucial aspect of flexibility.

That coordination is only possible if some information exchange is in place: DSOs have to gather information from all users connected to its networks and pass on the necessary data in an aggregated way to the TSO.

5.1.8. *Lessons learned from several international business cases*

This overview identifies innovative solutions for flexibility management by the DSO. Three geographical areas were studied: Germany, the United Kingdom and the USA (PJM[2], New York ISO).

2 PJM is an ISO in the states of New Jersey, Pennsylvania, Ohio, Virginia, West Virginia, Maryland, Delaware, and part of Wisconsin, Illinois, Michigan, Kentucky, Tennessee and North Carolina. The District of Columbia is also covered by the PJM.

The valorization of flexibility available to the distribution network is still under-developed and often integrated marginally into operation and flexibility programs at the scale of balancing plate (transmission grid). Few DSOs display important maturity in this regard.

5.1.8.1. *Case in Germany*

Local market (experimentation) – Energieversorgung Weser-Ems AG (EWE) (Utility y/c DSO): EWE set up local market system that allows generators to valorize their electricity on the local network, *via* hourly and quarter-hourly contracts with *Day Ahead*. Nevertheless, the majority of transactions are ultimately made with the national players.

Autoconsumption rates: an autoconsumption bonus existed in Germany until 2012, when it was halted. A "market integration" mechanism replaced the usual rates: generators of over 10 kW had to either resell at least 10% of their generation on the market or autoconsume it.

5.1.8.2. *Case in the UK*

Contractualization – Electricity North West (ENWL) (DSO): ENWL identified two primary 13 MVA substations under constraint that required investments (additional cooling equipment and change of transformers capacity to 23 MVA). A load shaving program (*via* the company EnerNOC) was established at the level of these two substations in order to reduce the load by 2–5 MW for 30 h per year, for 5 years. One of the lessons from this case is that *this remuneration is insufficient to efficiently activate flexibility sources*: EnerNOC struggles to constitute a load shaving stock in the context of this contract.

5.1.8.3. *Case in the USA*

Contractualization – PJM (USA)/Commonwealth Edison (ComEd), (Utility y/c DSO): ComEd developed load shaving programs in order to respond to two needs: control the peak and defer investments into identified primary substations. *ComEd can activate load shaving locally, all the way to the feeders* (400 A, 12 kV). ComEd offers very flexible contracts to customers over 1 MW (highly hour-seasonal rates), that go as far as to integrate flexibility contractualization for the largest customers. Incidentally, *ComEd* (which is also a supplier) *can make these contracts with customers who use a different supplier.*

Contractualization – New York (USA)/ComEd (Utility y/c DSO): ComEd divided its network into *70 load pockets* in the 60–250 MW range, which it can, *via* a request for load management resources, decrease load during emergency events and limit the peak on heat-wave days in order to *defer investments into primary substations* (up to 300 M $ for the most important ones).

5.2. Participation of end users to flexibility services

5.2.1. *Introduction*

Technically, flexibility used by DSOs could involve dispatchable generation, responsive consumption and reactive power regulation. According to EDSO analysis, the potential service providers are active users who have the capabilities of modifying their injection/consumption patterns: small industrial and commercial users, aggregated household customers and distributed energy resource (DER).

Small industrial and commercial users

Small industrial and commercial users could provide services to DSOs, either under the umbrella of an aggregator or individually. Small-medium enterprises (SMEs) using electricity intensive machinery, having a fleet of electric cars or using electricity for thermal control, such as air conditioning, water heaters, boilers, freezers or refrigeration, would be of particular interest, due to thermal inertia.

Household customers

Household customers could be an important source of flexibility for network operators, as long as providing flexibility is a transparent and effortless process. Individual household provision of system flexibility services is highly unlikely due to its weak impact on the system, however, household customers could potentially contribute to a pool of flexibility through their supplier or an aggregator. Several residential appliances, if pooled, could have an impact on grid management: electric vehicles, heat pumps, electrical heating and air conditioning are examples of devices that can provide flexibility.

Distributed energy resources (DERs)

A number of different DERs could be used to provide flexibility. Research projects, such as REserviceS or Improgress, have shown the potential of PV and wind, but other technologies, such as cogeneration, could be a resource worth tapping into. Whereas large DER units could act individually in flexible markets, small units, such as solar panels on a single house, will have to be represented by an aggregator to provide services.

DER controllability and forecasting will be key to make the most of its flexibility potential. If variable energy forecasting methods are quite reliable over a large territory, these techniques, today, lack the necessary accuracy to predict local energy patterns on day-before to intra-day timeframes. New smart prediction or other contingency tools will be needed in distribution management systems.

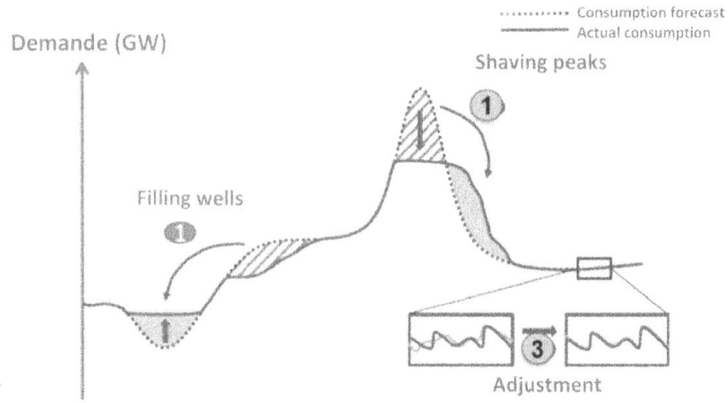

Figure 5.3. *Different uses for load shaving (source: Commission de Regulation de l'Energie, Smart Grid Paris 2014 presentation)*

5.2.2. *Focus on different tools and services downstream of the smart meter*

The smart meter will allow us to elevate the services related to electricity distribution to the same level as those prevalent in other sectors: banking, telecommunication, etc. Electricity market players will be able to go further in digitalizing their relation with the client.

5.2.2.1. *Services related to consumption data*

With more detailed consumption data readily available, the supplier will be able to offer its clients a series of information services, including a history of monthly consumption over the past 24 months, consumption over the current period in € and in kg of carbon dioxide (CO_2), maximum loads reached, etc. For a fee, the customer will benefit from more frequent information, alerts and personalized advice.

With this information, customers will have better control over their consumption profile and will be able to decide, when appropriate, to alter their consumption habits (for example Watt & Moi project in France).

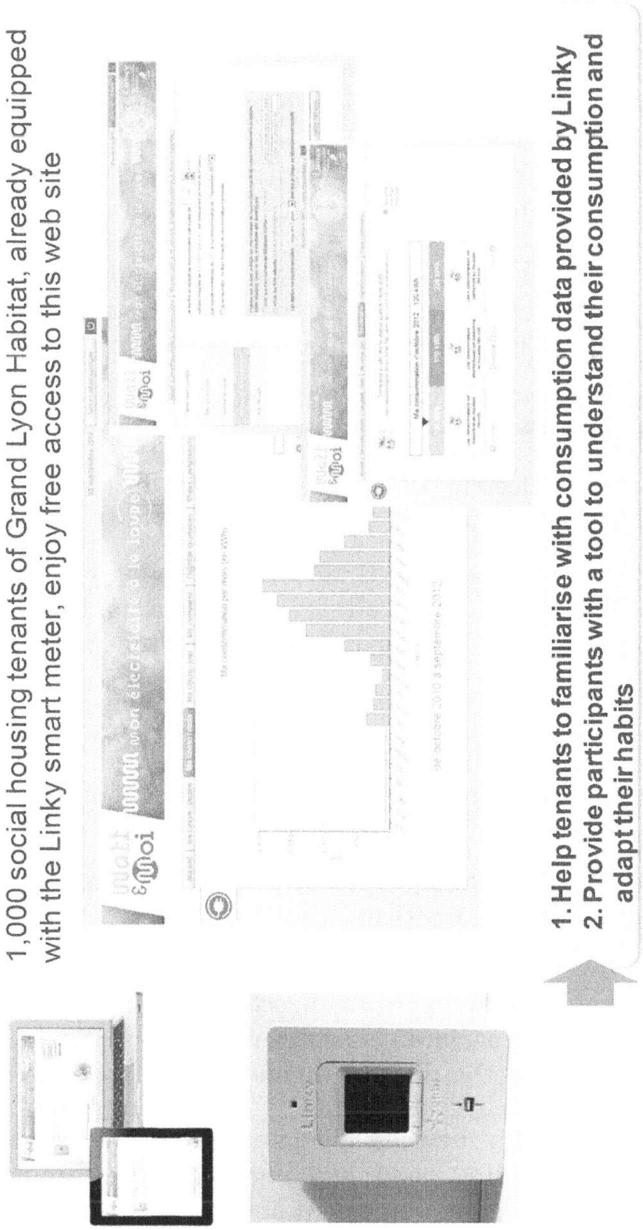

Figure 5.4. *French demonstrator Watt & Moi (sources: www.erdf.fr/medias/dossiers_presse/DP_ERDF_15032013.pdf)*

5.2.2.2. *Pricing offers – case in France*

The smart meter provides the possibility to record consumption on a calendar of 10 pricing ranges each day. For each pricing range, the smart meter allows associating a service for controlled usage, such as refilling hot water tanks or recharging electrical vehicles.

The supplier will be able to provide innovative pricing offers, within the limits of customer acceptability, for example, rates that encourage customers to:

– defer usage to outside of peak periods (e.g. 18—20 h for the Eco offer 18–20);

– charging electric vehicles at nightly rates (or following a favorable price signal);

– refill hot water tanks during sunny periods in areas where local solar photovoltaic production is higher than consumption (or also synchronized with wind energy availability).

5.2.2.3. *Load shaving offers*

The smart meter allows for two types of load shaving offers:

– shaving by pricing incentives, through moving peak calendars with day-1 notice;

– shaving with usage management (see Figure 5.5). Load management can be operated directly if the receiving device features an appropriate sensor to receive and interpret the signals, or indirectly, through an energy management device.

Complimentary devices downstream of the meter also allow for providing load shaving services, through, for example, an energy management device or a dedicated box.

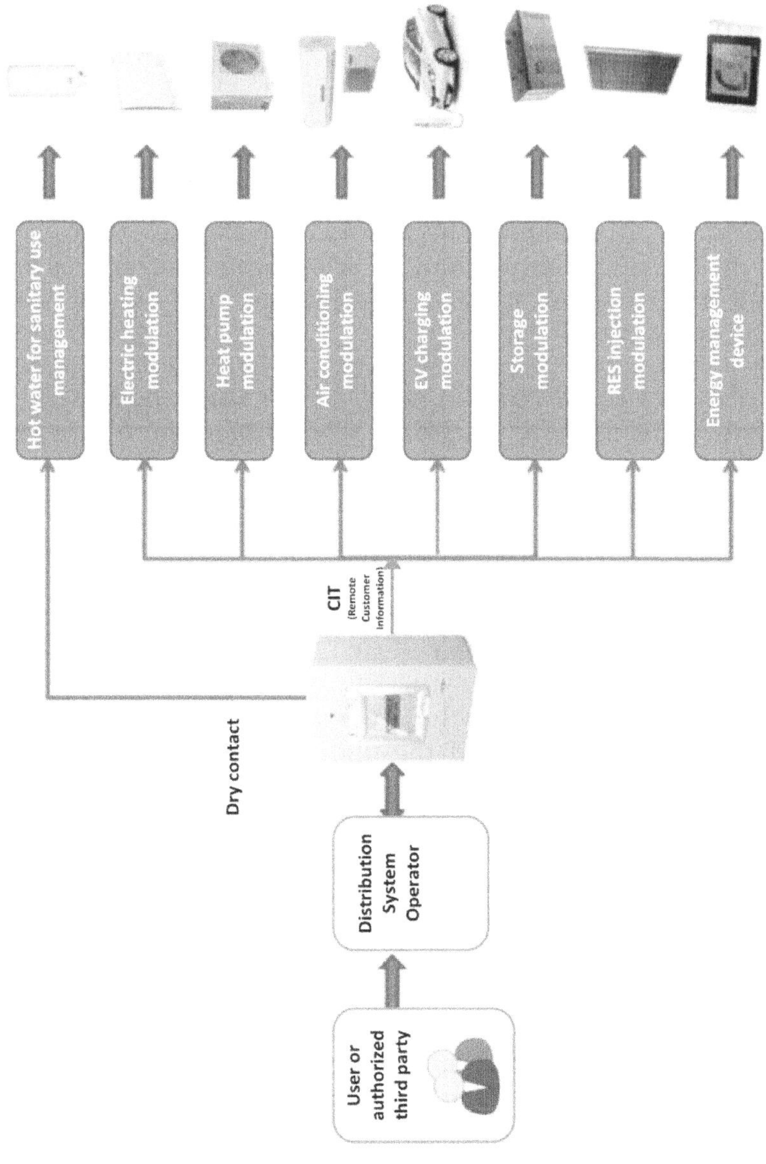

Figure 5.5. *Example of usage management via smart meter (source: [CRE 14c])*

5.2.2.4. *"Double sided market" offers*

"New" services that use consumption data, with customer consent, and presenting interests for other economic players, will also be able to develop.

The 'Two-Sided Market' economic model is the subject of university studies and mathematical modeling[3]. It is based on providing different functions or services to several parties and valorizing the positive externalities that exit between each of them.

The service offered to one of participating parties is an occasion to valorize usage or behavioral data by another party. This model is characterized by credit card payment platforms, where the "given" means of payment is also an opportunity to establish a source of behavioral data that are of great value to players, who seek to commercialize a product. In this model, a player can subsidize a service for the benefit of a third-party, if there is hope to make a profit out of the interaction with this third-party. Likewise, a service can be entirely offered to a player through an intermediary platform, if the data that it generates are of sufficient value to the other parties.

The first such services will bring in players such as insurers, solar panel maintenance companies or even telecommunication operators. They echo the fundamental preoccupations of the end-point customer: be alerted to electrical outages at the customer's home if the customer is absent, allow surveillance of solar panels installed by the customer or even be informed of incident management conducted by the customer's telecom operator during an asymmetric digital subscriber line (ADSL) outage.

The scope of these services will develop in the wake of the smart home/smart building, some examples of this include:

---

3 See [ROC 02].

– remote thermal diagnostic of the premises;

– troubleshooting and preventive maintenance of equipment;

– modulation of insurance premiums according to the equipment;

– preventive measures based on observed behavior.

5.2.3. *The necessary engagement of end-customers*

Demand-side flexibility is based on the assumption that consumers are willing to engage in demand-response activities. Engaging consumers will require incentives and technologies and is crucial for demand-side flexibility to work and deliver its full benefits:

– from basic passive techniques where the consumer has little to no control: the utility applies the solution without informing or consulting the customer and does not allow opt-in or opt-out capability (e.g. load shedding and voltage reduction);

– to more active techniques where the consumers take a hands-on role in determining the programs that they will participate in and the extent to which they are involved: the consumer can opt-in or opt-out at any time (e.g. time-of-use rate plans and utility controlled thermostats).

Several demonstrators across the world are working on the customer engagement issue. According to International Smart Grid Action Network (ISGAN)[4] in its latest case book on Demand Side Management (DSM), it appears that customers accept to change their habits but do not tolerate service degradation. Also in many cases, communication with the public tends to focus on the social values and

4 For the first CEM in 2010, ISGAN was launched and a multilateral cooperation agreement was created to favor the development of smart grids technologies and systems.

environmental aspects rather than individual financial benefits of participating in the project.

However, by participating in a demand response demonstrator, *the consumers often need to be guaranteed, from the beginning, that they will not lose money nor quality of supply*. Once the participant has some experience with the new service, if the participant perceives that the service has improved, his/her engagement should sustain.

Managing customer expectations is paramount in a long-term research or demonstration project. Both residential and commercial customers are risk averse and require simple and understandable products and services (plug and play).

Moreover, to tackle customer fears regarding data privacy issues, *clear rules on the ownership and protection of their data should be defined* and shared early into their engagement with the project.

5.2.4. *International benchmark and lessons learnt*

To tackle the different levels of maturity between countries, there are a number of coordinated multinational initiatives driving DSM research.

One of Europe's notable initiatives is through the EU FP7-funded project *Active Demand Value ANd Consumers Experiences Discovery (ADVANCED)*[5], which was launched in 2012 with the aim to better investigate behavioral barriers and fill in the gap in best practices availability for active demand (AD) design from the customer standpoint. In particular, a framework enabling residential, commercial and also industrial consumers to participate in AD will be developed, and impacts and benefits for key stakeholders will be quantified according to different scenarios. Real data from pilot projects underway in several countries –

5 www.advancedfp7.eu.

ADDRESS pilots in Spain and France, E-DeMa in Germany and Enel Info+ in Italy – will be used together with the data collected in VaasaETT's database from more than 100 European AD projects involving around 450,000 residential consumers. The ADVANCED project involves European Union (EU) energy utilities, universities, research centers and consulting firms in the energy sector, market research agency and one of the first aggregators in Europe.

Other worldwide initiatives arise through country cooperation. That includes not only cases from Europe, but also cases from other continents such as America and Asia. International Energy Agency (IEA) ISGAN is one example of these initiatives, where knowledge sharing is encouraged to:

– address standardization and interoperability, in order to improve business cases and assure the diffusion of the implemented solutions;

– develop cooperation between DSM actors/DSO/other players and provide the adequate services (e.g. continuous communication between the load and aggregator and between the aggregator, virtual power plants (VPPs) and system operator).

The latest ISGAN case book on DSM points out that *several obstacles are still to be overcome*. Indeed, the market design around the world is still "not mature".

Hereafter, four cases are summarized (see detailed presentation in section 6.3).

5.3. Data management as key success factor

5.3.1. *DSOs have a long experience in data management*

DSOs have a long experience in data management, collecting, validating, managing and providing data in order to manage their own activities in an efficient way, as well as giving support to market activities.

Case	Demand side management features	Lessons learned
The Netherlands – Power Matching City	– Smart devices automation (micro-CHP, heat pumps, electric vehicles, etc.) – Dynamic pricing – Virtual power plant	– Readily available technologies enable the implementation of a smart grid. – Significant reduction of imbalance costs (from 40 to 60%) caused by DER is achievable through the coupling to flexible clusters. – Need to design appliances, including house-hold appliances, differently.
Japan – Kitakyushu Smart Community Creation Project	– Dynamic pricing – Incentive program – Energy management systems	– Dynamic pricing system needs to be reviewed to implement an actual demand-supply balancing operation. – The customper is senstive to the frequency, time band and the method of notification/request sent to consumer. – The higher the price is, the greater is the effect on peak shaving.
Canada – A Virtual Power Plant To Balance Wind Energy	– Virtual power plant – Energy management System – Aggregator	– Communicating about environmental benefits of the project is effective. – Customers are sensitive to new types of relationships with utilities (e.g. cobranding).
Denmark – EcoGrid EU: Real-Time Market Demonstration	– Dynamic pricing – Customer responsiveness – Automation	– Customers are sensitive to the aspect of smart devices. – Communicating about social values of the project is effective. – The education of the local population regarding energy supply issues is key in the success of smart grid deployment.

Figure 5.6. *Several cases of the ISGAN case book about demand side management*

DSOs provide already a large range of data to consumers, market actors and public entities for various purposes: for consumer billing (in countries where metering activity belongs to DSOs), for settlement process in energy markets, for energy forecasting for TSOs, local authorities and to support DSOs processes (such as planning, connection and access and operation):

– *metering point administration*: DSOs manage a database of all electricity meters (metering point). This database contains information about the supply of electricity to each physical address. One important piece of information found on the database is the meter point reference number;

– *supplier switching*: all the necessary information is exchanged between the new and former supplier, and the DSO, with the latter acting as a neutral interface, checking and validating the switch;

– *consumer metering data for billing and customer awareness purpose*: DSOs facilitate a robust, efficient, non-discriminating and transparent electricity market by providing information for customer awareness and market stakeholders. They store and provide consumer metering data;

– *allocation and reconciliation*: following the energy market opening, DSOs have endorsed responsibilities for aggregation and extrapolation of complex data for the allocation and reconciliation process. Therefore, DSOs have developed strong skills upon collecting data, aggregating and modeling production profile as well as consumption patterns;

– *settlement*: accurate and timely data from smart metering could significantly improve the quality of settlement, which in turn could encourage innovation and efficient use of energy.

Such activities are operated under regulatory supervision and guided by cost efficiency principles, based on well-established processes and market transactions, and making use of the best available information and communications technology (ICT) and solutions from reference vendors and service providers.

DSOs, who have developed skills and know-how of managing suppliers' perimeters, can handle the activity to manage flexibility perimeters and ensuring non-discriminatory access. Indeed, financial flow and the outcome of transactions are based on these perimeters. In the meantime, DSOs have to locate these capacities to ensure the safety and security of network distribution.

5.3.2. *DSO, the market facilitator*

Increased data flow means more opportunity to make use of data, and the need to develop new and flexible technical solutions, to deal with this flow of data, while guaranteeing data security and consumers' privacy[6].

Markets involving capacities and flexibilities are going to require increasingly treatment on the metering data. For instance, in the particular case of demand response, energy cannot, by definition, be measured. It can only be estimated by comparing the amount of energy that would have been consumed at the exact period of time if no action would have been triggered, the baseline, and the physical consumption load curve.

DSOs have a key role to play, as explained by EDSO, in managing and providing data to the market in particular for perimeter handling and settlement in new market

6 With a measurement every 10 minutes for 10 million customers, there is a total of 1.5 billion meter readings per day. Considering a ratio of 400 octets per reading, the daily volume of data to manage is about 580 Goctets.

mechanisms. Having DSOs involved in the settlement phase and perimeter handling benefits all parties. With its position within the system, DSOs guarantee reliable and fair markets within the system. Over the last 10 years, DSOs have proven their ability to meet these objectives.

5.3.2.1. *Being a market facilitator*

In the near future, the fundamental role and responsibilities of DSOs regarding data management will fundamentally be the same as today: being a market facilitator, allowing access and connection in a non-discriminatory and transparent way and taking care of security of supply and quality of service. But as technology and network users are quickly evolving, DSOs roles and responsibilities should be able to adapt.

DSOs already have technical expertise to manage data: each DSO, depending on the applicable regulation, will take decision on which technology to deploy, how the data will be transferred and how IT companies may support them on this task.

DSOs are the best-suited contender to perform efficient and non-discriminatory data handling to enhance the functioning of the market and consumers' participation.

6

Pilot Projects and Use Cases

6.1. A global dynamic with regional specificities

The previous chapters brought to light the stakes of the distribution network and the necessity to integrate advanced functions to accompany the emergence of intermittent renewable energies, electric vehicle (EV) and charging station use, services for more flexible demand management, etc.

The majority of these advanced functions do not exist on an industrial level. Before proceeding to a wide-scale deployment of these functions, a period of experimentation under real life conditions is necessary to analyze and to select the solutions that provide the most value to all the players of the electric system. Several dimensions need to be studied:

– *technical validity*: on the one side, ensuring that the different technological solutions provide the answers expected by distribution system operators (DSOs) and, on the other, develop interoperability between the building blocks from various areas (electrical engineering, IT, telecommunication…) and from different manufacturers;

– *economic model*: analyze the implementation costs of smart grids functions, identify and quantify the benefits for all the players of the value chain; ultimately determine the possible business models and the capacity to finance the investments of a generalized deployment;

– *adequacy of regulation:* ensure that the evolution of the role and the responsibilities of the actors of the value chain do not present a risk to the security of the electric system; identify the funding models (and the level of return on capitals) that allow the DSOs to cover the risk of deploying advanced technological solutions (higher risk than deploying more robust hardware and equipment such as cables and transformers);

– *replicability:* assess the capacity of the solutions and the designs to adapt to different technical and regulatory contexts; and ultimately contribute to their "standardization" (or at least their adoption by several DSOs/several countries) to reduce the costs through a more important scale effect;

– *client acceptability:* include the consumers into the experiments on demand management and flexibility activation; understand their motivations and analyze their uptake of various services and mechanisms related to consumption shift or demand reduction; estimate the capacity to efficiently mobilize (in time and space, guaranteeing the required response level) flexibilities on a large scale, going from several hundred to several thousand customers.

Main initiatives	North America (DOE)	Europe (JRC)	Asia (various sources)
No. of projects	>100	459	>350
Cumulative budget	8 Bn $	3.2 Bn €	>10 Bn $

Table 6.1. *Numerous smart grids demonstrator projects are undertaken in several regions of the world*

However, the great variety of contexts leads each of these regions to use its own experimental approach and focus on specific technological solutions:

– emerging countries face strong growth in electric consumption and experience a phase of rapid development of their generation, transmission and distribution infrastructures;

– North American and European countries have aging infrastructures that need to be renewed, while keeping in tune with the development of low carbon emission energies and favoring energy efficiency;

– certain Asian countries (Japan, Korea, etc.) seek to reduce their energy dependency on fossil fuels, as well as greatly reduce their greenhouse gas emissions.

A common point to the majority of these demonstrators is the role played by the DSOs: as well to support the investment effort of the R&D as to ensure the standardization and integration of building blocks and to encourage the cooperation between the different parties involved.

6.2. North America

6.2.1. *Drivers of smart grids development*

In North America, the main objective is to create a more efficient electric system by: detecting incidents faster, modernizing older equipment, reducing frauds and phasing out manual reading of consumption indices.

The primary trigger of the development of smart grids was the black-out of August 2003 in the Northeastern United States. This event, which deprived 55 million people of electricity and cost the economy 6 Bn $, brought to light the aging state of the facilities and the ineptitude of electrical networks in managing large-scale incidents. A Task Force then recommended, in 2004, the deployment of new technologies to modernize the network infrastructures and increase their reliability.

The level of necessary investments to integrate these recommendations is very high and the existence of over 3,000 facilities further fragments the dynamic. The EPRI estimated that the cost of a complete transition of the US network to smart grids will be 476 Bn $ over 20 years. This important trigger of the first investments was the American Recovery and Reinvestment Act (ARRA) of 2009, which allocated a budget of 4.5 Bn $ specifically for the R&D of smart grids.

The development of renewable energies is another driving force behind the development of smart grids, especially since the creation of the Renewable Portfolio Standards (RSP) that established, in 29 American states, a minimum percentage of renewable energy sources (RES) generation (in MWh) or a minimum installed load (MW).

In Canada, the primary driving force of smart grids development is the development of programs, related to demand management and activation of flexibilities, to reinforce energy reliability and efficiency.

6.2.2. *Primary experimental approaches*

In the United States, the smart grids Act is supported by the US Department of Energy and is composed of two main programs (Figure 6.1):

– A program of investment subsidies in electric distribution and transmission networks (the Smart Grid Investment Grant – SGIG). There are 99 projects with a total budget of 8 Bn $, of which 3.4 billion are subsidized by the federal government. At the end of May 2014, 1.8 Bn $

worth of equipment and software were deployed in distribution networks with the SGIG[1].

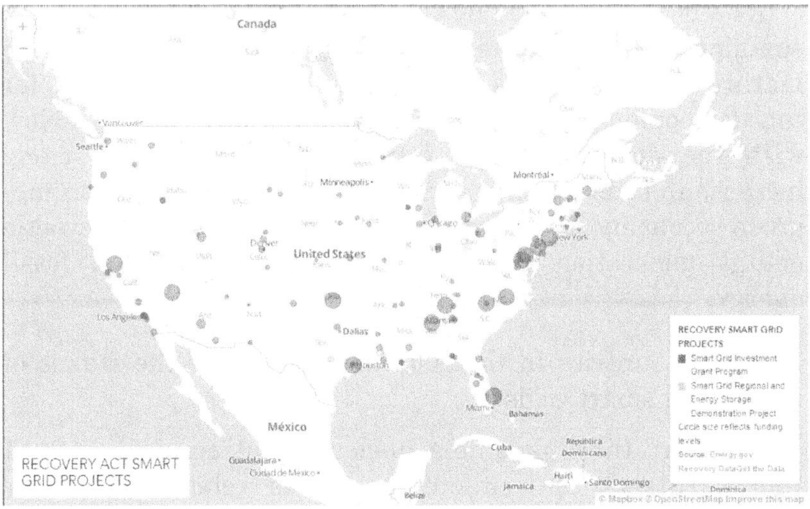

Figure 6.1. *Map of smart grids projects participating in Recovery Act*[2]. *For a color version of the figure, see www.iste.co.uk/boillot/smartgrids.zip*

A program of demonstrator projects will test innovative concepts and technologies. Two types of demonstrator projects are considered: "regional"-scale demonstrator projects (sensors, networking, failure detection and energy flow management) and projects dedicated to the storage of energy (batteries, compressed air, flywheel). There are 32 projects selected by the DOE for a total budget of 1.6 Bn $, of which 600 million are subsidized by the federal government.

1 Source: https://www.smartgrid.gov/recovery_act/deployment_status/distribution.
2 Source: US Department of Energy: http://www.energy.gov/maps/recovery-act-smart-grid-projects.

6.3. Asia

6.3.1. *Drivers of smart grids development*

Asia is characterized by the combination of countries with sustained economic development and countries with a more mature economy. This provides a double opportunity for the development of smart grids: on the one side, to follow with the development of infrastructures and the strong increase in consumption and, on the other, to transition into green economy to increase the proportion of renewable energy generation and reduce the dependency on fossil energy.

Three countries in the Asia area illustrate the stakes of developing smart grids:

1) Japan*:* the Fukushima nuclear plant accident of 2011 was a catalyst in reinforcing actions related to energy efficiency and accelerating the deployment of smart meters. Moreover, Japan succeeded in decreasing CO_2 emissions and in the development of renewable energies, a major objective. This is manifested in an ambitious project to equip over 5 million houses with solar photovoltaic (PV) panels and therefore reach a capacity of 28 GW by 2020[3];

2) South Korea: it has also defined an ambitious plan to increase the proportion of renewable energies in the energy mix, with the goal of reaching 11% in 2030, as opposed to 2% in 2009[4]. Hence, the sustained growth of the industrial sector's demand for electricity (~5% per year and >50% of the total) is also an important factor in the development of smart grid infrastructures and technologies;

3 Source: NEDO – an introduction to Japan's "Smart Community".

4 Korea Smart Grid Institute (KSGI): Korea's Smart Grid Roadmap 2030.

3) China: the rapid development of its economy and the population's standard of living constitute the primary driving force for the development of new infrastructures and smart grid technologies: development of transmission grids and distribution networks, integration of renewable energies, development of electric mobility, urbanization and the emergence of Smart Cities. The investments foreseen by the State Grid Corporation of China (SGCC) in the electric networks are in the order of 40 Bn $ per year, of which 10–15% are in smart grids technologies and equipment[5].

6.3.2. *A proactive experimental approach*

China and South Korea put into motion their experimental measures in the context of a more global approach, planned over 15 to 20 years from pilot projects, to industrial deployment. These approaches are simplified by a strong dynamic propelled by public authorities and the presence of public electricity enterprises, in a position of monopoly, covering the entirety of the value chain.

Hence, SGCC (88% of the territory, 1 billion people served) is engaged in a three-step program, which partly corresponds to the "12th five-year plan" of the Chinese government:

– preparation (2009–2010): elaboration of an action plan and launch of the R&D program;

– construction (2010–2015): conducting pilot projects, fine tuning advanced technologies and investing into technologies and hardware;

5 Zpryme Research & Consulting, LLC – Smart Grid Insight: SGCC Profile – March 2012.

– refinement (2016–2020): consolidation of technologies and unification of infrastructures to establish a “smart grid”.

The level of investments is at the scale of China. Thus, SGCC foresees a global investment of 100 Bn $ into smart grids throughout the 3 periods, of which 9.2 billion are used during the first one of 2009–2010[6]. There are also 298 demonstrator projects which will be conducted until 2015 in 26 provinces. 300 million smart meters will also be deployed by the end of 2015.

For its part, the Korea Electric Power Corporation – KEPCO (>90% customers) – participates in the Smart Grid plan of action established by the Ministry of Knowledge Economy, notably in charge of the energy sector:

– large-scale pilot (2010–2012): implementation and operation of a pilot smart grid to test new technologies;

– extension of experimentations (2013–2020): extension of pilots in metropolitan areas and implementation of smart solutions in customers’ homes;

– large-scale deployment (2021–2030): setting up a smart grid on the scale of the entire country.

In this context, KEPCO foresees an investment of 7.2 Bn $ by 2030[7], of which 400 million is for the first phase and 800 million is for the second. The primary experimental measures are concentrated on Juju Island (~6000 accommodations) around 5 topics (Smart Place, Smart Transport, Smart Renewable, Smart Power Grid, Smart Electricity Services). Here, 18 consortia are mobilized and regroup a total of 171 enterprises.

6 Zpryme Research & Consulting, LLC: Smart Grid Insight China.
7 Reuters http://www.reuters.com/article/2011/02/18/kepco-korea-idAFTOE71H00R20110218.

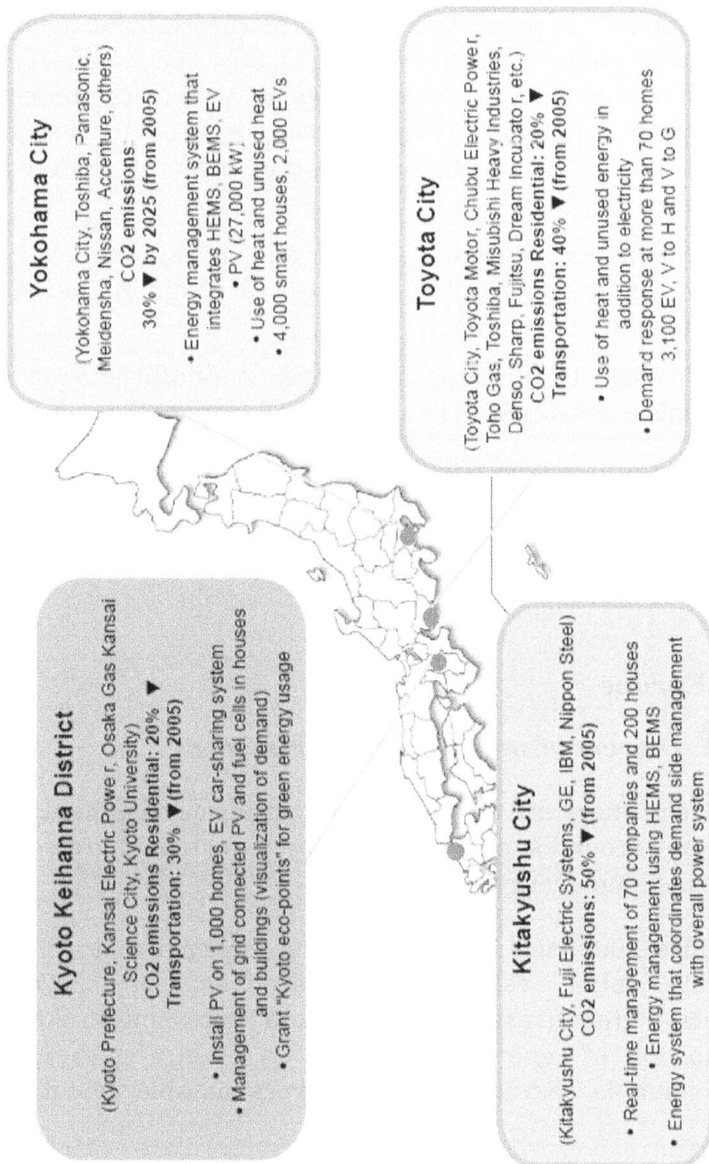

Figure 6.2. *Introduction to Japan's "Smart Community". Source: New Energy and Industrial Technology Development Organization (NEDO)*

In Japan, stopping the nuclear power stations following the earthquake of 2011 brought up the resilience of infrastructure as a major challenge, as well as the necessity to develop energy sources that can be used in case of a catastrophe. The four major programs[8] are each centered on their own objective:

– Yokohama: introduction of an energy management system on the scale of buildings and communities (4,000 accommodations);

– Toyota city: increase the energy self-sufficiency of new residential constructions and "demand response" programs;

– Keihanna: encouraging rates for current inhabitants (700 habitations);

– Kita-Kyushu: dynamic pricing and management of a private network.

6.4. Europe

6.4.1. *Drivers of smart grids development*

The primary driver behind smart grids development is the dynamic around supply security, the reduction of greenhouse emission and energy efficiency.

Since 2008, the European Union engaged in a "3 × 20" climate plan establishing objectives for 2020 of reducing greenhouse gas by 20% and increasing to 20% the proportion of renewable energies in energy consumption. This plan is characterized by a very notable evolution of

8 Note that since 2011, Ministry of Economy, Trade and Industry (METI) also finances 8 smaller-scale technological demonstrators of similar subjects (demand response – Mie University campus, energy management in a Nagasaki parking, etc.).

the European energy mix with a shift away from fossil energy generation means and the massive development of intermittent renewable energies (wind and solar PV).

Installed capacity (GW) / Type of RES	In 2008	To date	Target 2020
Wind	65	105 by end of 2012	201
Solar	10	78 by end of 2013	90

Table 6.2. *Installed and targeted RES capacity in Europe. (Sources: European Commission – Energy infrastructure, EurObserv'ER and Eurostat barometers)*

These two energies are characterized by important daily and seasonal variations. The networks that these energies are fed into have to be equipped with the new smart grid technologies in order to maintain a permanent balance between generation and consumption.

In January 2014, the Commission published a proposition in favor of the energy climate policy for 2030, which is notably characterized by the goal to reach at least 27% of renewable energies in the final energy consumption.

A second driver behind smart grid development originates from the 3rd energy package that was adopted in 2009 and established the goal to deploy smart meters at no less than 80% of consumer homes by 2020.

Finally, the development of interconnections between countries of the European Union and the renewal of aging infrastructures constitute a third important driver behind smart grid development.

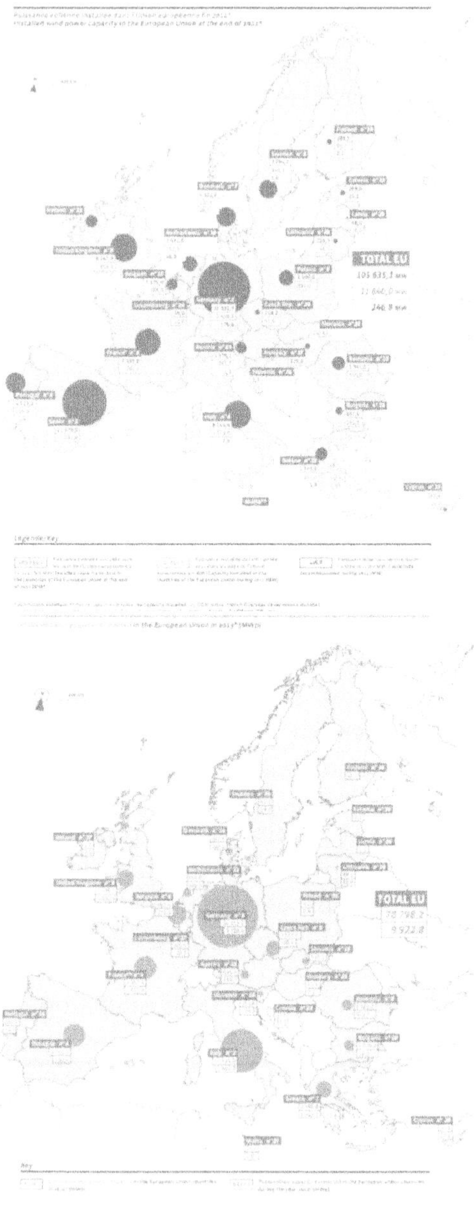

Figure 6.3. *Solar and wind power installed in Europe (Source: EurObserv'ER 2013 and 2014 barometers – http://www.energies-renouvelables.org). For a color version of the figure, see www.iste.co.uk/boillot/smartgrids.zip*

6.4.2. *Primary experimental approaches*

There are many smart grid demonstrator projects at the heart of European Union members. According to the Joint Research Center's annual report of May 2014 (Joint Research Centre (JRC) Smart Grid Projects Outlook 2014[9]), 459 projects were launched since 2002 and represent an investment total of 3.15 Bn €. The most dynamic countries are France, United Kingdom, Germany, Spain and Italy, which represent over 60% of the total investment. *The DSOs are the most invested players in these projects.*

There are two types of these projects: the projects led at the heart of each country and the multinational projects. They cover the various issues of smart grids, centered around 6 topics: smart network operation (primary topic), solutions downstream of the meter, aggregation of flexibilities (consumption and generation), integration of EV charging infrastructures, integration of RES generation and smart metering.

The first feedback shows that the primary barriers are not necessarily technological. There are two types of major challenges. The first concerns the regulatory context and the flow that still exist in many countries between *the allocation of roles and responsibilities, the sharing of the costs and benefits and ultimately the new business models.* The second concerns consumer resistance and their engagement in different tests run throughout Europe.

The method of financing these experimental measures is a determining factor in the European dynamic. The operation of electric networks is an activity regulated by the majority and the investment decisions are closely related to pricing structures defined by the national regulators.

9 http://ses.jrc.ec.europa.eu/sites/ses.jrc.ec.europa.eu/files/u24/2014/report/ld- na-26609-en-n_smart_grid_projects_outlook_2014_-_online.pdf.

Therefore, the support of the European Union played a large role in initiating Smart Grids projects, with support programs such as FP7 (50 Bn €, of which 2.3 were consecrated to energy), set up during the 2007–2013 period, and soon the Horizon 2020 program over the 2014–2020 period.

6.5. The European project Grid4EU, fosters and accelerates experience sharing

6.5.1. *A large-scale demonstration project bringing together six European DSOs*

Initiated in the context of a call for FP7 projects by the European Commission, the project Grid4EU, coordinated by the European Regional Development Fund (ERDF), unites six European electricity DSOs in order to prepare the development of the electric networks of tomorrow.

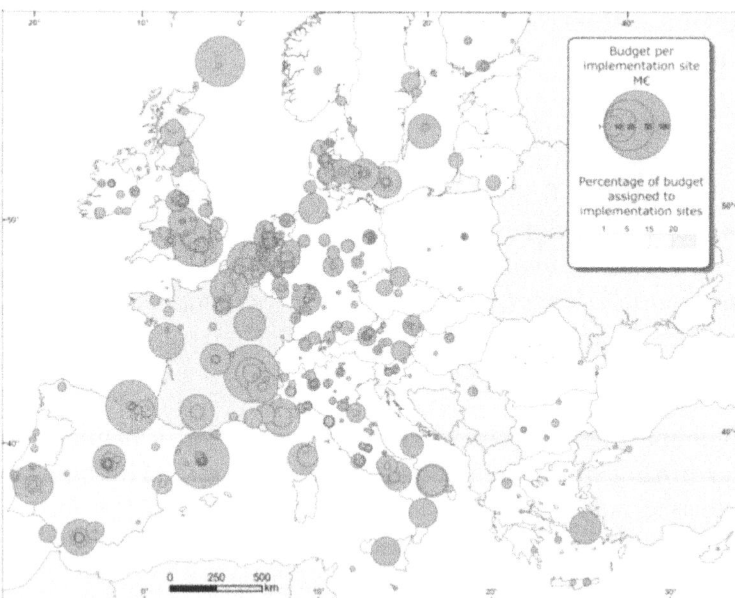

Figure 6.4. *Map of smart grid projects identified by the JRC in 2014. For a color version of the figure, see www.iste.co.uk / boillot / smartgrids.zip*

Several objectives set by the EC:

– develop and test innovative technologies;

– assess the scalability of new experimental technologies;

– guarantee the compatibility of solutions between the different European contexts;

– make standards emerge through demonstrators;

– analyze costs and benefits of smart grids (business case).

Six complementary demonstrators:

This project involves the establishment of six demonstrators in different European countries:

– *locations:* Reken (Germany), Uppsala (Sweden), Castellon (Spain), Forli-Cesena (Italy), Vrchlabi (Czech Republic), Carros (France);

– *total budget:* 54 million euros (financed up to 25 million euros by the European Commission).

The project addresses many topics related to smart grids: consumption management solution, energy storage, localized islanding, supervision functions and operating medium voltage (MV) and low voltage (LV) networks, controlling energy consumption peak management programs etc.

A coordination to favor of sharing experience and competences

The central goal of this project is to guarantee the coordination between different demonstrators. Therefore, this program favors the efficiency of an approach divided between different projects, specifically with transversal research actions and shared results between the various DSOs involved, in particular based on the “business use cases” method.

GRID4EU also takes advantage of the competences of other industrial and scientific partnerships; the project federates around 30 partnerships from 12 countries of the European Union.

6.5.2. *DEMO 1 (Germany – RWE) MV network operation automation and determining the ratio of decentralized intelligence in secondary substations*

6.5.2.1. *Project description*

Efficient due to a significant RES penetration rate, RWE wishes to enhance MV network monitoring and operation.

Currently, RWE does not have a tool to control the MV network. Thus, the objective is to implement a system based on local automation for operating MV networks with a coupling between centralized and decentralized operation.

This demo aims to install local automations on critical points in the network, capable of communicating between themselves and the different equipment of the network. *The objective is to know the state of the MV network from the data collected and to make decisions locally.*

The decisions taken locally will be conveyed to the centralized control center, which carries out global monitoring of the network and can intervene on the decisions made by controlling the automations remotely.

6.5.2.2. *Specificities of the context and key values*

Demo 1 is located in Germany, in the Reken commune of the North Rhine-Westphalia region. Choosing Reken (145 secondary substations, 120 km of MV lines) is mostly due to the strong presence of solar PV and wind generation (32.3 MW connected) and a near-equilibrium between generation capacity and maximum network load (25.5 MVA for 14,000 customers).

6.5.2.3. *Specific objectives*

– Testing advanced MV network operation functions;

– obtaining a clear picture of the state of the network in real time;

– allowing local automation to optimize the topology of the network locally.

6.5.2.4. *Use cases tested by the demonstration*

Implementation of decentralized automation on the MV network: when the load flow is not optimal, the operational scheme is modified to re-optimize it and reduce losses.

Management of MV failures: gather alarms, transmit them to supervisory control and data acquisition (SCADA), and automatically conduct recovery operations to recover the highest number of customers.

Monitoring and remote controlling of the MV network: real-time knowledge of the state and the load of the network, the location, the status and the operational state of equipment, transmission of reports of automatic operations conducted on the network, possibility to deactivate automation when required.

6.5.3. *DEMO 2 (Sweden – Vattenfal): a tool for LV operation and in particular identifying LV failures*

6.5.3.1. *Description of the project*

The experimentation aims to study the opportunity to implement network services based on the advanced metering infrastructure (AMI) infrastructure (LV-PLC or low voltage power line communication).

It is a matter of collecting information from the smart meters and the RTUs located in secondary substations to implement a tool for LV network supervision.

6.5.3.2. *Specificities of the context*

The Vattenfal demonstrator is located in the town of Uppsala, 50 km north of Stockholm. This zone presents the particularity of acting simultaneously as a rural and an urban network. The demonstrator counts over 145 public distribution (PD) substations for around 15,000 customers. Here, the introduction of smart meters is finalized and AMI systems have been operational for several years.

6.5.3.3. *Specific objectives*

Detection of outages on the LV network involve gathering outage information, reporting to the operator, calculating outage duration at local and national levels, and informing the customers.

Improving the follow-up of supply quality would involve:

– collection of data on LV feeders: voltage value levels, active and reactive power, hourly peak, load curve, transformer alarms (overload, temperature, etc.);

– collection of customer data: voltage levels, active and reactive power, hourly peak, load curve.

6.5.4. *DEMO 3 (Spain – Iberdrola) MV and LV failure detection, reconfiguration of the MV network during an incident*

6.5.4.1. *Description of the project*

The Spanish demonstrator aims to define the best technical infrastructure capable of simultaneously improving supervision of LV and MV networks, and quality of supply. The primary developments concern:

– secondary substations: collection of MV and LV data with the help of RTUs, failure detectors, measurements (voltage, current flow, power, transformer temperature, etc.) and data issued by meters and concentrators. Implementation of communication between sensors and centralized tools via use of MV power line communication (MV PLC) to transmit all collected data;

– information system: enhancement of algorithms and data collection processes, and implementation of a LV supervision tool.

6.5.4.2. *Specificities of the context*

The Iberdrola demonstrator is located in Castellon, Spain, in the Valence region (a town of 200,000 inhabitants), equipped with 20 secondary substations and smart meters.

6.5.4.3. *Specific objectives*

– Use the information supplied by the smart meters to gain better understanding of the state of the LV networks (LV failure detection, voltage and voltage quality excursions);

– monitoring from the distribution substation with sensors/RTU in the substation and in the LV lines, to assess overloads, unbalancing, etc.;

– assess losses (technical and non-technical) by comparing the curve of the distribution substation with the customer total, hour by hour;

– reinforce network operation with the implementation of an MV failure detection system with appropriate recovery schemes, which allows for rapid recovery of a maximum of clients.

6.5.4.4. *Primary use cases tested by the demonstrator*

Secondary substation of the future and MV and LV infrastructure management: supervision of the LV network (failure detection, lack of voltage, neutral point rupture, phase imbalance, etc.), detection of MV and LV over/under-voltages, remote control of MV breakers, monitoring transformers, detection of technical and non-technical losses.

Automatic Outage Detection: automatic detection of outages through analysis of data received from different equipment (substations, lines, meters).

Automatic Grid Recovery: automatic detection of failures, alarms, dynamic marking of network sectors affected by the incident, automatic recovery of customers.

6.5.5. *DEMO 4 (Italy – ENEL) economic model and technical operation of storage, MV voltage regulation, anti-islanding of decentralized generation*

6.5.5.1. *Description of the project*

The primary objective of the Italian demonstrator is to improve the capacity of the MV network to integrate decentralized generation means, in particular solar PV, by using active control of the generators connected to the MV network, the control of the load on the MV network and the storage.

6.5.5.2. *Specificities of the context*

The demonstrator is located in the Emilia-Romagna region in the Forli-Cesena province in Italy. It is a rural zone with high RES integration, specifically solar PV (40 MW, with 24 generators of over 500 KW connected to the MV network) and weak consumption.

In particular, the network participating in the demonstration is composed of 2 primary substations with 20 MV feeders, 160 secondary substations and at least 5 PV generators.

6.5.5.3. *Specific objectives*

To reach these objectives, the demonstrator:

– implements algorithms for regulating voltage on MV network;

– manages the reconnection of RES from the network in case of unintentional islanding;

– tests the use of storage for the optimization of network management (1 MW/1 Mh).

6.5.5.4. *Primary use cases tested by the demonstrator*

– *Control and supervision of voltage on the MV network:* calculate voltage profile and send appropriate instructions to the RES, the storage and the load regulators in the transformers (on load tap changers);

– *anti-islanding protection on MV networks:* alerting all RES generators present on a MV feeder following the opening of a top circuit breaker, in order to synchronize their disconnection;

– *load shedding for MV customers*: in agreement with the customers, activation of interruptible load shedders (77 active MV customers for 40 MW).

6.5.6. *DEMO 5 (Czech Republic – CEZ) operating islanding with co-generation, MV and LV failure detection and reconfiguration of the MV network following an incident*

6.5.6.1. *Description of the project*

This demonstrator is part of a larger project called "Smart Region", at the heart of the Vrchlabi distribution network, in the Czech Republic.

This project implements:

– automation of MV and LV networks with the help of equipment remotely operated *via* an infrastructure of telecommunications and smart meters, and active customer participation;

– automatic operation of islanded areas by ensuring acceptable supply quality throughout the duration of the islanding.

6.5.6.2. *Specificities of the context*

This demonstrator is located in the North-East of the Czech Republic, in the Liščí Kopec region (semi-urban area), in the town of Vrchlabi (5,000 customers, 13 km of MV lines, 65 km of LV lines, 30 secondary substations).

6.5.6.3. *Specific objectives*

– Improve MV and LV failure detection on the network and automate the modification of automatic recovery operation schemes to reduce outage time on the MV and LV networks and resupply the most clients in the least time;

– ensure the operation in islanded mode of an area with a source of local generation provided by the combined heat and power (CHP) unit by guaranteeing local balance and sufficient supply quality with the help of smart meters.

6.5.6.4. *Primary use cases tested by the demonstrator*

Automation of failure management on MV and LV networks: detection and isolation of failures by the SCADA, implementation of alternative operational schemes, automatic return to normal operational schemes after repairs.

Management of islanding functions: study of automatic disconnection of a part of the network while maintaining generation/consumption equilibrium with the CHP.

6.5.7. *DEMO 6 (France – ERDF): project NiceGrid*

6.5.7.1. *Description of the project*

NiceGrid tests solutions that contribute to *simplifying the insertion of solar PV electricity generation into LV networks*, by optimizing electricity generation, consumption and storage at the scale of a district of the Carros commune.

6.5.7.2. *Specificities of the context*

The project is developed at Carros, in the plain of Var, at the French Riviera Nice Metropolitan area and within the scale of the "Eco-Vallée" National Interest Operation.

The department of Alpes-Maritimes is located at the extremity of the transmission grid, in an area fragile in terms of electric supply, which features important sources of renewable energy, most notably solar.[10]

6.5.7.3. *Specific objectives*

– Develop and test the energy management system to optimize the generation/consumption balance at the scale of a district;

– ensure control of LV network load in the state of *massive injection* of photovoltaic electricity generation;

– give users an active role in managing their electricity and test the behavioral response of consumers/generators;

– study and test the techno-economic interest of implementing electricity storage distributed between

10 End of March 2014, PV generation installed capacity is 1040 MW with a waiting list of 640MW.

different levels (the primary substation, secondary substation and final customer);

– experiment with islanding on a restricted pocket of the network.

6.5.7.4. *Primary use cases tested by the demonstrator*

Optimization of local balance in the presence of PV: supervision of the LV network (consumption and generation forecast, simulation of the electrical state of the network, detection of constraints, arbitrage between the available flexibility levers), optimization of MV and LV network operation, controlling storage and RES.

Power shedding: control of the power consumed by the customers (activation of flexibilities and transmission of pricing signals), validation of third-party shedding programs.

Customer information/DSM: management of data from meters and communicating consumption information.

6.6. An approach based on use cases

6.6.1. *Definition*

The methodology of "use cases" was developed to equip smart grids projects in regard to the *description of the business processes to create and the IT functions that support these changes.* The objective is to simplify their insertion in the IT blueprint and to simplify the developments of associated IT.

The DSO defines its strategic intentions and turns them into "use cases", i.e. scenarios on a given subject (for example, optimization of RES connections, forecast management of RES generation or voltage setting). The necessary equipment and developments are then set up: sensors, softwares, communication technologies, etc.; and objectives (KPI, or Key Performance Indicator) are specified, for instance, the "rate of sensor availability must be over 90%".

This is a method that begins with a strategic framework, but grows on the feedback issued from the projects. To avoid adopting a short-sighted strategy based solely on technology, it is necessary to begin from the goals, field testing scenarios and consolidating results to construct the overall design.

6.6.2. *Advantages*

This method presents numerous advantages for structuring the capitalization of the feedback issued from smart grids projects, feeding industrial projects, anticipating the impacts on the work-flow and the competences of the DSOs, and driving the reflexive process behind operational changes. It also provides a basis for discussions between the numerous external players concerned by these projects. It enables mapping with precision both tested and untested "smart services", in order to opportunely orient future projects.

6.6.3. *The development of use cases*

Pioneer utilities, such as Southern California Edison (SCE), and other R&D institutes, such as the Electric Power Research Institute (EPRI), developed and use the use cases methodology. EPRI created a methodology guide in 2010, in the context of the "Intelligrid" program, for the use of smart grid project directors who would like to operationally implement this approach.

In Europe, the International Electrotechnical Commission (IEC) decided to promote the "Intelligrid" method as a reference in order to standardize it[11].

This method is well adapted to smart grids as it offers:

– a universally comprehensible formalism that allows us to represent and to model the complexity of the smart grids "system of systems", in a methodological context that is simultaneously flexible and structuring;

– a pragmatic process of collective description that includes many of the participating parties (managers, working and IT professionals, partnerships, etc.), specifically adapted to the context of R&D projects and smart grids demonstrators;

– a coherent and structured multi-level description that enables the analysis of challenges from different points of view strategic, operational and IT).

The capitalization of Use Cases of Smart Grids projects and demonstrators must facilitate the integration of their results into the evolutions of businesses (IT directing scheme, evolutions of network materials and systems, etc.) and the overall transformation project.

The implementations of the smart grid use cases are numerous:

– for internal use, to consolidate a detailed up-to-date picture of smart grids, in order to refine strategic positions, axis of R&D, etc;

---

11 Processes launched during the creation of the *Ad Hoc* Group 5 at the Technical Committee 8 (TC8), later picked up by Working Group 5 (WG5) of TC8 created in Oslo in November 2012. The first version of the standard IEC 63559-2, related to the model of textually describing a Use Case (Use Case template), was published in 2013.

– for demonstrators and smart grids projects, to respond to the need for project specifications and to promote the vision of smart grids to participating parties;

– for Europe, via European projects, in order to promote the DSOs' point of view on the evolution of their position and their business model;

– for international standardization, in order to promote the vision of the evolution of electric systems operation.

The use cases offer the DSOs a possibility to enter a cycle of formal description of their business processes. Today, the evolutions of business design methods and techniques enable the consideration of significant optimizations in the coupling between the description of business processes and the implementation of their associated IT via direct control over the exchange infrastructures.

6.7. Focus on some advanced projects of the ISGAN case book about Demand Side Management

At the international level, the European Commission participates in the Clean Energy Ministerial (CEM). The CEM unites 23 governmental structures. Its goal is to support the development of technologies for clean energy and, in general, support the transition into a sustainable energy system.

For the first CEM in 2010, the International Smart Grid Action Network (ISGAN) was launched, creating a multilateral cooperation mechanism that favors the development and the deployment of smart grids technologies and Systems.

The participating governments launched in 2011 an agreement for implementation under the supervision of the

International Energy Agency (IEA). This agreement grants ISGAN a governing body that notably allows the participation of players from the private sector.

Case book context

On the occasion of the fifth Clean Energy Ministerial, the ISGAN case book called "Spotlight on Demand Side Management"[12] has been officially released.

The case book includes studies from 12 countries: Austria, Canada, Denmark, France, Germany, Italy, Japan, South Korea, South Africa, Sweden, the Netherlands and the United States, covering a broad range of contexts: economic, political, geographical, structural, cultural and market.

The lessons learned and best practices provide qualitative insights into the complexity involved in deploying Demand Side Management (DSM) initiatives.

The 12 case studies are based on a diverse range of technologies and are carried out under specific market rules. They incorporate various programs and policy mechanisms, and include information on the costs and the associated business cases for investment.

There is a number of best practices and common themes emerging from these 12 case studies that is likely to be useful to any stakeholder investigating or deploying DSM.

The following sections focus on the four most advanced projects of the ISGAN case books.

12 Source: http://www.iea-isgan.org/?c=5/112/367&uid=1300.

6.7.1. *Denmark – EcoGrid EU*

The fundamental principle of the EcoGrid EU real-time market concept is to exploit flexibility in electricity consumption. At residential level, flexibility is in this case typically associated with heat pumps and electric heating either for hot water or for keeping the house warm.

6.7.1.1. *Objectives*

The objective of the EcoGrid EU project is to illustrate that information and communications technology (ICT), and innovative market solutions can enable an operation of a power system with more than 50% RES such as wind, biomass and photovoltaic (PV).

6.7.1.2. *Design*

The EcoGrid EU market concept is based on the publication of real-time price signals. Adapting the behavior of flexible resources like electric heating and heat pumps will contribute to maintaining the balance of supply and demand in the power system.

The test households and the commercial customers are organized in groups depending on the 'price control system' they belong to: 500 households in the manual control group, 1,200 households in the automatic control group and up to 100 industry/commercial buildings with automatic price control.

6.7.1.3. *Lessons learned*

The recruitment process is considered a success although various delays in the project have been a challenge. During the entire recruitment process the communication with the public was focusing on the social values and environmental aspects; the participants are guaranteed that they will not 'lose money' by participating in EcoGrid EU.

The local electricity company in Bornholm, Østkraft, expects that energy will be a very important part of the customer service in a smart grid deployment scenario. The European consumer organization (BEUC) in their analysis of 6 recent studies on the use of smart meters *reveals that the actual savings average 2–4% in the best cases, where consumers have clearly opted for their use.*

6.7.2. *Japan – Kitakyushu Smart Community Creation Project*

To achieve the objective of efficient energy usage, the following actions are implemented:

– establishment of a new energy system centered in a community energy conservation site;

– DSM of customer participation type such as "dynamic pricing" and the "incentive program".

6.7.2.1. *Design*

The Community Energy Management System (CEMS) is the core of the field demonstration project, installed in the Smart Community Center. It communicates with Homes, Buildings, Factories, and Retail *via* AMI. In the first year of the demonstration project, a new, variable critical peak pricing scheme that set five levels of peak price was proposed. The goal was to have residents participating in the project respond to peak prices and to ascertain what price level, if any, is effective depending on how urgent the demand–supply situation of electric power may be.

6.7.2.2. *Results*

In summer, the price of electricity during peak hours was set from 15 yen for level 1 to 150 yen for level 5. Residents were notified a day ahead of time as to what the peak price would be.

Levels 2 to 5 were charged 10 days each, a total of 40 days, in the summer of 2012, when the highest temperature exceeded 30°C.

The peak shaving effect of the treatment group ranged from about 9–13%, indicating that the higher the pricing level is, the greater the effect.

6.7.2.3. *Lessons learned and best practices*

About 85% of 230 general households and almost 100% of the companies in the Higashida district participated in this demonstration project.

The high participation rates were attributed to multiple outreach efforts by city hall explaining the project to business owners and general residents and requesting their participation. The residents and offices did not have to pay any expenses to participate in the project.

6.7.2.4. *Method*

In these experiments, the targeted households have been randomly divided into two groups; "treatment group" (dynamic pricing is applied) and "control group" (dynamic pricing is not applied). This method, called Randomized Control Trial, is based on the guideline of US Department of Energy (DOE). Comparing these two groups, it is possible to verify the effect of the demand response with the dynamic pricing.

6.7.3. *The Netherlands – PowerMatchingCity*

PowerMatchingCity is a live lab demonstration in the Netherlands with, among others, the following objectives:

– demonstrate an integral optimization method based on local markets and distributed intelligence for both capacity and commodity;

– extend the role of the grid operator: validate the peak load reduction potential by extending the trial with households to the transformer.

6.7.3.1. *Design*

The project was launched in the village of Hoogkerk, in Groningen, in 2009: PowerMatchingCity I, with 22 households, hybrid water pumps, solar PV systems, "smart" devices and 2 EVs. This project was completed successfully and is now receiving follow-on stimuli from the Dutch Government: PowerMatchingCity II, with an additional 18 households, 10 EVs and 2 smart distribution transformers.

6.7.3.2. *Results*

The trial showed that it is possible to create a smart grid or energy network with the associated market model using existing technologies. The system enables consumers to exchange electricity freely and the level of comfort is maintained.

By using the PowerMatcher, more renewable energy may be integrated in the electricity system; it was shown that approximately 65–90% of the off-shore wind power, which would normally not be used without coordination, could be locally utilized.

It is technically feasible to allow demand response to track supply: Measurements from the micro-CHP, the hybrid pumps and the charging of EVs all indicate that the system responds quickly to fluctuating demand.

6.7.3.3. *Lessons learned and best practices*

Smart appliances

There is a need to design current appliances, including 'smart' house-hold appliances. Appliances should be able to automatically switch on or off depending on the current electricity rate. This implies that devices should be programmed to access, via the internet, current electricity rates.

Automation

The consumer acceptance level is high, and a clear change in 'energy behavior' can be observed. The effect on their direct energy consumption is limited, but the end-users show an increased willingness to invest in more energy efficient appliances as soon as their investments result in the expected savings.

Customer satisfaction

Households get their bills every month, based on their actual electricity consumption. Two types of contracts have been developed, after extensive consultation with the participants: cheapest energy bill or maximum energy produced locally.

6.7.4. *Canada – a virtual power plant to balance wind energy*

PowerShift Atlantic is demonstrating fully grid-integrated virtual power plants (VPPs), designed to allow for more effective integration of wind power.

6.7.4.1. *Objectives*

The primary objective of this demonstration is to determine if load shifting can provide an economic and effective alternative to building new supply side ancillary services for the integration of wind with minimal to no disruption to participating utility customers.

6.7.4.2. *Current status*

To demonstrate value as an ancillary service there must be a significant volume of load available for shifting the VPP via aggregators. Customer participation is therefore essential to the success of this project. A new level of trust must be established in the relationship between the customer and the utility in order to allow the utility to shift loads.

By March 2014, the project had approximately 18 MW of controllable load through a combination of commercial and residential customers. Overall customer satisfaction with the program remains high (80%), with an expressed desire for more frequent detailed program information.

6.7.4.3. *Lessons learned*

The consortium decided to develop an innovative VPP solution, to provide load shape management and to provide the equivalent of a 10 min spinning reserve on demand.

The VPP operates continuously in near real-time, therefore continuous communication between the load and the aggregator, and between the aggregator, and the system operator, is required.

PowerShift Atlantic utility companies have found that the project has enabled them to build new types of relationships with their customers that foster new opportunities and methods for communication and involvement. Despite not having financial incentives or other guaranteed monetary benefits, customers were keen to participate in order to contribute to finding ways of integrating wind energy more efficiently.

Commercial customers tend to have lower tolerance for risks from interruptions compared to residential customers. To mitigate any risk associated with this, customer support services have been reviewed and employees interacting with the customers have been trained to achieve a high level of expertise on the project.

7

Smart Grids are the Future for DSO

7.1. Advanced smart grids for DSOs worldwide

7.1.1. *The evolution towards smart grids is ineluctable*

A large part of the distribution systems operators (DSOs) in the world is confronted with the effects of the energy transition. The evolution toward smart grids is inevitable. Prior to engaging in the latter, it is essential to investigate the general state of the current electric system. To rationalize the approach, a framework is necessary in the form of an investment master plan into the networks. Without such measures, the DSOs face the risk of forgetting the final overall goals by adopting an approach solely based on technology and the short term.

In certain emerging countries, the priority could be set on basic electrical and technical equipment. In Europe, the objective is to integrate renewable energy source (RES) at the lowest possible cost, and therefore defer as much as possible the investments required by network reinforcements. In the case of France, it is crucial to master the peak. One way to reach this goal involves the capacity market and load management.

Information technology (IT) and telecom designs need to be made the subject of an evolutionary master plan. The master plan allows to guarantee technical coherence and to search for the economic optimum. However, technical evolutions are very fast, and feedback from smart grid projects need to be integrated "as they come" into the DSO's decisions. The "use cases" methodology should be used to achieve this. The latter is a bottom-up approach that results in translating the DSO's strategic intentions into concrete objectives related to equipment (sensors, telecom solutions, software, etc.).

In emerging countries, it is important to implement network automation as soon as possible. Indeed, this becomes more complicated if not done from the start. The modernization of networks has a double objective: reinforce the "strong current" components and insert a "light current" component, which amounts to simultaneously managing the development of electric networks and IT/telecom networks.

Smart grids only gain economic substance when there are real synergies between these two domains. Here is an illustration through several examples:

– it is necessary to develop a telecom network to anticipate the installation and the usage of remote operation devices on the electric network;

– smart metering with an appropriate telecom solution (for example, power line carrier (PLC) in South Europe) is essential for the decrease of non-technical losses;

– without an IT master plan, there is the risk of disoptimization, leading to the decrease of the functional capacities of different systems.

The basic techno-economic calculation is a good tool to valorize invested capital. The challenge is to structure the network to evolve toward performance. The DSO defines a

"Merit Order", which allows the classification of technical choices and the definition of priorities: either to reinforce the network or to favor intelligence and automation. Externalities (employment, environment, etc.) may also be integrated into the calculation.

Often, there is a lack of consideration toward good network structures through lack of competence and tools. For instance, when equipment goes out of order, rather than launching a heavy project of "asset management", it is preferable to find solutions that are simple answers to simple technical problems. This allows defining and reaching quality targets quicker.

This requires an industrial designer who is not a stakeholder in the investments. Cross-division work turns out to be essential to integrate the technological choices into the business.

7.1.2. *The development of smart grids is a necessity for the DSOs*

Failure to anticipate the development of these "smart" technologies may introduce risks to the DSOs:

– the DSO will not be a major actor in the electric system, leading to a subsequent re-examination of the DSO economic model;

– interposition of new players ("local energy system operators") between the local authorities, customers and DSO, hence taking a part of the value of the DSO's activity, currently "socialized";

– risks for the protection of people and equipment in the presence of decentralized generation on the network;

– failure to master the effects of load shaving on the network;

– over-sizing works for managing constraints that only appear punctually, with a disoptimization of investments in reinforcement and connection.

Although various studies have already been undertaken on different smart grid topics, generally, starting form a problem type (photovoltaic (PV) or electric vehicle (EV) integration, etc.), a conceptual solution (load shaving or reactive power management) or a family of components (smart meter or storage equipment), many stages remain to be covered before selecting the most pertinent tools amid the multitude of available options:

First, a key element will be experimentation associated with situating certain objects in a system approach.

Second, beyond the usual components of the network (electrical and technical components, electric distribution structures), there will be development of "active" components: decentralized means of generation, means of storage, dynamic and modifiable loads at customer homes *via* load shaving mechanisms. The laws governing the behavior of these objects and players depend in part on external stimuli (pricing signals and orders from the suppliers or aggregators). They remain to be established through experimentation. Measuring the response of these entities and actors in real conditions, specifically the customer acceptability of these new interactions with the network, is one of the important stakes of the smart grid demonstrators and pilot projects.

Third, the economic value of the different solutions and the conditions under which this value is expressed, in terms of adaptation of the regulatory context[1], needs to be assessed.

1 If a solution is economically viable for the collective, how should the rules be adapted to make this solution emerge? If a solution is not economically viable for the collective, how to ensure that the rules will not facilitate its emergence?

Once these stages are passed (tests of individual solutions, system tests, behavioral tests of the users of the system, economic and regulatory selection), it will be necessary to integrate the solutions that will be ultimately kept in the DSO's processes (operation/development).

7.1.3. *But also an opportunity*

– *Maintain a hi-tech image*, with technological innovations for both the outside (customers and communities) and inside (evolution and enhancement of jobs and competences of the employees);

– *position large DSOs as key players in the evolution of the network*, on the one side, by their capacity to launch research and development (R&D) programs inaccessible to players of a smaller caliber, and on the other side, by their neutrality and search for economical optimum for the collective through their regulated activity;

– *respond to new expectations of customers and market players* (supply of new data, adaptation of access conditions, etc.).

Moreover, the scenarios of energy policies that the DSO will have to face are quite varied and depend on factors related to strong uncertainties (growth rate, price of energy, substitution from thermal uses toward electricity, regulatory models, etc.). *The additional flexibility brought to distribution networks by smart grids, complementing the range of solutions that the DSO can implement, will increase the ability of DSOs to adapt to these evolutions.*

The distribution network undergoes changes that are difficult, but progressive. The deployment of technological solutions on the network will likely follow the rate of apparition of new constraints and solutions, that is on a timescale of the order of 10 years. The alteration of the end-point user's behavior in relation to their energy consumption and their power needs will also take many years, as it is suggested by the feedback from pilot projects. Smart grids are in total continuity with the existing distribution networks.

7.2. A necessary evolution of skills and jobs of the DSOs

7.2.1. *Competences are necessary to conduct experimentations successfully and to get the most feedback from them*

In order to experiment on smart grid technologies and to conduct cost-benefit studies, DSOs take part in numerous demonstrations in Europe (over a hundred). These projects mobilize important resources at the heart of the DSOs:

– at the regional level where the experiments take place: project management teams, operating teams (usage, primary substations, operation, mastery of decision and execution infrastructures);

– at the level of the managerial team: global direction of the program, support from technical experts, contributions from information and telecom systems specialists, etc.

The competences used are essentially abilities to manage complex projects, due to the variety of the domains concerned (network, information systems, economic models, etc.) and to the diversity of the interlocutors (industrials, local authorities, energy mastery agencies, European Commission and research laboratories).

7.2.2. *Once the experiments are finished, the resources and competences need to be reinforced in preparation for large-scale industrialization and deployment*

It is too early to precisely establish a plan for introducing smart grid functionalities on the network. Nevertheless, some guidelines are emerging. The deployment will begin by outfitting the network with the necessary equipment for advanced management and functionalities that will be deployed afterward.

This outfitting will concern primary substations (control of operation, sensors and automations) and the medium-voltage (MV) network (sensors, automations and devices for transmitting data, generally at the level of the remote control device (RCD) and secondary substations). The outfitting of the low-voltage (LV) network will essentially take place *via* the smart meter. A function "measuring" the global load can be incorporated at the level of secondary substations.

The reinforcement of project management, information and telecom system competences is necessary during the deployment operations. A site for data (network and customer) is to be foreseen upstream of the deployment of smart grid functions (completeness and quality of references).

Finally, the solicitations of local authorities in terms of smart grids, eco-neighborhoods and electric vehicle charging infrastructures, are increasingly numerous. DSOs will have to acquire project leaders, experts in smart grid technologies and concepts, to guide the local authorities to enact the chosen evolutions in the context of the energy transition.

New jobs and competences will need to be perpetuated for the operation and maintenance of electric networks of the future.

The mastery of smart grid technologies and functionalities requires having new competences in the following five domains:

1) *Making forecasts*: consumption and generation forecast scenarios, at different geographical and time-scales.

2) *Operating telecom and information system infrastructures*: development of telecom infrastructures concurring to the development of the electric network, supervision/operation/maintenance of telecom networks, administration/management of telecoms and equipment (software versions, cyber-security, supervision/operation/ maintenance of smart equipment (metering systems, automations and network sensors).

3) *Operation of local electric systems*: analysis/forecast of constraints on the network, forecast management in relation to planning and operation, take into account load shaving programs, activation of flexibility levers to lift network constraints.

4) *Management of relations between the players of the energy system*: personalized connection solutions, management and follow-up of the contractual relation, balance sheets, improvement cycles, interface with players of the energy system (transmission system operators (TSOs), aggregators, suppliers and customers).

5) *Management of large volumes of data*: formatting, treatment, publication, integration of market mechanisms, protection and updating.

The fundamentals of the "core jobs" competences will be enlarged: the complexity of the technologies and competences will dictate, on the one hand, reinforcing high-level technical expertise and, on the other hand, obtaining the most automated IT tools and processes for the DSO. The interfaces between jobs will evolve greatly, in particular dynamic management of constraints will lead to higher cross-over between the areas of "study/planning" and "operation/usage". The development of local energy systems that allow optimizing the network at the local scale entails a need for reinforced coordination between MV and LV networks.

The deployment of more intelligent and more communicative equipment and systems will lead to a stronger penetration of IT competences, in particular in the area of "operation/usage", and render data management even more strategic. The relation between the players of the energy system will increase and require the reinforcement of contract management competences.

The development of services around consumption data (publication, reporting and generation of reports) will require competences in data management, administration and protection, as well as high-level expertise of statistical analysis. The requests of local authorities will become more numerous, in particular to supply data on energy systems of the territories: competences for data management (that, which can be supplied, which is part of the contract, which is specific), aggregation and analysis of data.

7.3. The French electrical sector mobilizes: the "Smart Grids" plan[2]

Today, France has certain advantages that allow it to play the leading part in the smart electric networks of tomorrow. With a strong industry among the world leaders in the

2 Source: New Industrial France: "Smart electric network" guideline, validated on 7 May 2014 at the Elysée Palace.

domains of electrotechnics, networking equipment and information system integration, France can also take advantage of leading network operators, R&D that is expressed through a large number of demos. This high-level research is led both by world renowned universities and by efficient R&D centers. This picture encourages ambition in the shortest delays. Indeed, smart electric networks are no longer the subject of laboratories, and numerous solutions are ready technically and technologically to become widespread.

In order to consolidate the French electrical and information sectors on these new markets, which are growing fast and creating employment, to improve their visibility and place French industries in the leading position of smart electric networks, three major axes need to develop:

– *in the short term, build "the French team of smart electric networks"* by federating the industry around a "label" and an operational structure with the means to ensure promotion and coordination between players, and hence gain access to a larger business volume, than through individual actions of each player;

– *in the medium term, transition from demos to targeted deployment* of smart grids solutions in a geographic area in France where the concentration of different actions in the industrial, research and educational fields will create a levering effect favorable to the creation of an employment pool, with a mesh of innovating businesses, to ultimately create economic conditions appropriate for the deployment and to promote the advantages and attractiveness of the industry;

– *in the long term, prepare the cost efficiency of the industry by 2020*, by refining the R&D strategy of smart grids, favoring the emergence of solutions essentially originating from small-medium enterprises (SMEs) and start-ups, by providing an "Open Innovation" platform, operators integration and industry support.

In support of these three axes:

– in terms of smart grids standardization, the efficiency of the French action will be reinforced *via* the implementation of a process at the heart of the French administrative instances to initially formulate, and then carry the French positions, and specifically, *via* a reinforced presence on the strategic field of interoperability standards;

– works will be carried out on the methodological tools, enabling the valorization of consequences of the smart grids deployment in terms of creating employment and value for the collectivity, while minimizing the environmental footprint;

– by leaning on the existing networks of "centers of competitivity", a guiding body will be organized to help SMEs, which carry innovating ideas in the domain of smart grids, develop at the heart of French and international markets.

8

Key Findings

8.1. Smart grids or the real network revolution

From the beginning, electric networks function along the same principles: they carry electricity from the upstream (power plants) to the downstream (customers spread throughout the territory). The system provides satisfaction: with this typical "branched" architecture, where cables branch into increasingly smaller diameters, performance is efficient both in terms of the cost and the quality of product and customer service.

The energy transition will profoundly alter the energy landscape with an important evolution of network constraints and the emergence of new markets in order to manage those constraints to the best (in France, the consumption peak grows twice as fast as the consumption).

The massive arrival of renewable energy sources (RES) and new uses (e.g. electric vehicles and heat pump) in the distribution network topples the traditional network operation models. Power flows in two directions: from the upstream to the downstream, when the centralized generation dominates, and from the downstream to the upstream, when the wind or the sun allows decentralized RES generation.

This evolution requires adapting the network: the issue is no longer sizing the network according to the relatively foreseeable unidirectional distribution. It is necessary to take into account the local level of variable bidirectional flows to develop and use the network at the techno-economical optimum for the collective.

To develop the ability to predict and influence energy flows, it is essential to reinforce the coordination between the players of the system, to find a balance between reinforcing networks and dynamically managing constraints, and to process large volumes of data and information in order to determine and activate the most efficient levers.

Network reinforcement often follows the development of RES in order to ensure their integration under conditions that guarantee security. The level of investments is as important as the distance between the location of RES generation and the consumption areas.

Furthermore, if they are too far from each other, it will be necessary to invest into allowing the export of generation surplus according to the demand of neighboring regions. In certain cases, this may require the creation of a new network, as the location of generation is located in unserved areas.

8.1.1. *Smart grids*

Smart grids represent the existing and future power network. They are not subject to technical revolution. The major challenge lies in the modernization and renewal of facilities. The distribution system operators (DSOs) will implement solutions that will allow using facilities to their full capacity (life expectancy, losses, etc.). *The communication network is fundamental* to smart grids: it enables connecting in near-real-time multiple sensors to the central system.

The DSO uses dedicated lines, power line carrier (PLC), radio, relay links or optic fiber. *Finally, the processing and monitoring system* integrates the latest developments in information technology. It must find a balance between centralization and decentralization, aim for performance despite the volume of data and tend toward the optimal integration level both functionally and geographically.

8.2. More RES means more network

The intermittent nature of RES generation requires the network to be able to ensure supply at full power to customers despite the wind and sun regimes. *The networks play the role of "mutual insurance", which guarantees the customers supply that is reliable* and secure in spite of the hazards related to the weather or consumption fluctuations.

The role of the DSO becomes pivotal in the distribution of tasks at the heart of the overall electric system: while the TSO has to ensure the balance between supply and demand on the national level and the stability of frequency, *the DSO ensures the continuity of voltage level at the local scale.*

The excess of RES generation raises the voltage levels. Conversely, too much demand results in lower voltage. This phenomenon can occur, for instance, during the concurrent charging of an important number of electric vehicles from the same line (along the same road), if the network sizing was not intended for such loads.

The regulations in effect allow functioning within a 10% deviation of the nominal voltage at the level of low-voltage (LV) networks. Outside of these limits, customer devices as well as network equipment can suffer damages in the case of excessive voltage elevation, while disruptions in customer devices and the general operation of the network can occur during strong voltage reductions.

Network voltage plans need to be revised in order to allow more flexibility for the operational modes, given the appearance of the new constraints.

The DSOs will act as *neutral market facilitators* and hence call upon flexibility services at different timescales in addition to the network reinforcement options.

Forecast management becomes a real job: software is developed to predict wind and solar photovoltaic (PV) generation according to weather forecasts. Likewise, software is developed to anticipate demand at the local scale. It is therefore necessary to foresee congestions (network constraints) that will appear at different locations and time frames.

The DSO will then seek to implement flexibilities to remove these constraints and guarantee stable operation of the network. The DSO develops a technical aggregation platform that facilitates to act at the offer level (RES generation), the demand level (charging and shedding) or the storage facility level, if the latter becomes competitive. It defines a "merit order", which will order different possible flexibility solutions according to their cost/benefit ratio.

8.3. The DSO is a facilitator

On its own, the DSO is not an actor of shedding, storage or generation. *It is neutral in relation to the market players and makes possible (facilitates)* the emergence and the implementation of new flexibility processes. It does not sell commercial services in competition with the market players.

The European network codes aim to create a common framework for all the flexibility actors. If necessary, the market rules (*intra-day and day ahead*) will need to be adjusted consequentially.

The generalized deployment of smart meters provides multiple advantages: first, for the customer, it makes possible billing according to a real index ("I pay what I consume") and remote operation (switching on/off, capacity shifting) conducted remotely and without bothering the customer. Moreover, in case of failure, it facilitates the diagnosis and accelerates repairs.

In France, with Linky, additional functionalities are planned to simplify customer information (access to an Internet site with data that is updated daily with access to previous consumptions the possibility to connect a radio transceiver to access real-time information). Linky is also able to act on house appliances through eight signals, given pricing tables received from suppliers.

Linky can operate not only on the hot water tank according to the peak/off-peak rates, as it already has been the case in France for 40 years, but also on other cables from the electric panel of the house: the electric vehicle (EV) charging terminal, electrical heating supply of each room or household appliances, for example. New tables, with a moving peak, will appear, allowing, for example, to charge EV during periods when solar generation is abundant at the meridian peak of summer or when there is wind (see "wash with the wind" programs in Germany).

8.4. Consumer or "consum'player"?

The residential customer will be better informed about his consumption. He will be able to become, *if he wishes so, a "consum'player"*, i.e. *he will influence his energy consumption profile by altering his behavior*. He will be informed about power demand peaks that can threaten the balance of the distribution network and potentially lead to a "blackout" on the local level (neighborhood, village).

This information will allow him to defer his consumption or to renounce on certain uses (putting off certain consumptions, slight reduction of heating temperature in certain rooms in the house for short and cyclical periods, defer launching house appliances) at certain periods of the day, week or year.

By analogy to traffic signalization, we can imagine three levels of intervention. *If the light is green*, the customer can decide on his own, from the elements of information available to him, to alter his consumption and peak. If the light is *yellow*, the energy supplier or any other player previously authorized by the customer can send his pricing signals, essentially influencing the customer's household appliances. The light becomes *red* when there is a threat of a "blackout". The DSO then has to act in the common interest: partial or total power shedding can turn out to be essential to allow a larger number to keep a minimum electrical supply.

Other customers–consumers will become generators by installing PV panels on their roof. The auto-consumption is acceptable as long as the financing of distribution networks continues to be ensured at a correct level by the carrying rates shared by all of the customers, including the generators.

Other customers may want to use the batteries of their own EV, cogeneration facilities or other local generation means.

8.5. Smart meter at the service of smart grids

Smart metering also aims to allow the DSO to observe the LV network in order to operate it better. Complexity is increasingly important for the operator and it is vital to be able to regard the LV network as the medium voltage (MV) network in order to employ an integrated approach and common tools to operate a single network. The information incoming from smart meters can be directed at the secondary substations to regional dispatching, for example.

In certain countries, considerable efforts are made to modernize and automate the MV network. In France, for example, the last 15 years saw the creation of 30 regional dispatch centers, the installation of 100,000 remote control devices (RCDs), the evolution of overall architectures by reinforcing electrical "loops", the implementation of efficient software for automatic "self-healing" (automated operation functions (AOF)) and the optimization of the operational scheme according to network constraints (automated network functions (ANF)).

8.6. A smart bubble?

There are still many obstacles to overcome in order to conduct this industrial revolution in the networks.

The necessary investments into smart grids require a fair distribution between the players. In most countries, the smart grid infrastructure is implemented by the DSOs. They benefit from it as do the other actors: RES generators, energy suppliers, aggregators, etc. How can the fair repartition of costs between the beneficiaries be ensured? Which incentives should be set up? How should the overall system be regulated?

The industrials say that they are ready, certain technologies are mature and the first tests took place with the pilot projects. *Only life-sized experience at the regional scale can confirm the hypothesis selected in terms of reliability and cost.* The DSO will, in particular, have to be vigilant in relation to generic defects, which are liable to degrade the reliability and the safety of the overall system.

Data management at this scale is a new challenge for the DSO, which will have to be ready to manage the considerable complexity and increase of the information volume. Numerous “Big Data” players see here an opportunity for growth; however, *it is essential for the DSO to keep control of this area at the risk of being deprived of a vital part of their mission.* They prepare actively for it: their level of competence and mastery of the system evolves positively. The challenges around cyber-security are handled with the most advanced technological solutions in order to fight the threat of large-scale failures and malicious attacks.

The industrial development will only be able to take place in the context of sufficient interoperability that would favor real competition between equipment suppliers and therefore lower the costs for the DSO. The standardization will need to be pushed to the European and the worldwide level to give Smart Grid technologies a universal characteristic.

The interface between the meters and the household appliances is subject to many technological choices (choice of communication bands and protocols), which makes difficult, or even impossible, the industrial deployment of devices for operating household equipment under economic conditions that are acceptable for most customers. *The viability of the economic model of DSM in the “residential” area remains to be shown.* This pitfall is related to the acceptation of new technologies and services by the end-point customer. It also depends on the fair repartition of the value created by the system with fragmented, but interdependent responsibilities.

8.7. Invest to save?

It is, in fact, essential from the get-go that the customer can benefit from new services offered by the meter (information, appliance operation, shedding, etc.). The Smart Building and Smart City projects will be at the origin of large-scale marketing operations, without guaranteeing the technical and economic feasibility of demand management.

A "Smart Bubble" is a genuine risk. An implementation failure can have a negative effect on the customers, who could reject the network modernization entirely, including the smart meters and all other intrusions into their private life.

The demand management is presented as a domain where energy suppliers and aggregators will be able to differentiate themselves by providing increasingly sophisticated offers. The consumer associations are opposed to the idea where customers subscribe to paying offers from the suppliers in order to make energy economies. *Despite the ongoing experiments and the pilot projects, the technical feasibility of devices for controlling household appliances and large-scale residential shedding remains to be proved.*

To make large-scale shedding of individual customers possible, complex algorithms still need to be developed. Furthermore, the method used for measuring shedding does not always fall under a consensus, although the panels method suggested by ERDF has already undergone the certification process by an independent recognized body, the GENEX-INSEE.

8.8. Smart grids: a genuine industrial opportunity

The electric network undergoes a profound evolution. The deployment of technical solutions takes place at the rate of the apparition of constraints related to RES and EV over a decade. *The smart grids represent a genuine industrial opportunity: many technologies are mature, many innovations are incoming,* the industry is motivated, demos are ongoing in many countries and they lead to the accumulation of rich feedback, the political world is interested by the concept of smart grids insofar as it appears successful in attaining its stated objectives in energy policies and important investments are necessary for numerous networks. All this constitutes an opportune moment to begin the deployment of new technological levels.

The expected benefit of smart grids will be easier to obtain if *the outfitting is made globally at the scale of the area or the region.* This is why important renewal (in Europe) and the development of a new network (in emerging countries) are key opportunities to launch the large-scale deployments. They will allow us to test coherently all aspects of smart grids: from network performance to customer service.

Smart grids will reinforce spectacularly the attractiveness of the DSO in relation to young engineers and technicians by mobilizing intellectual resources essential for the success of this transformation. Around the Smart Grids projects conducted by the DSOs, there are the beginnings of a complete ecosystem with start-ups, laboratories, universities, engineering schools and large industrial enterprises.

Bibliography

[ADE 13] ADEME, ATEE, DGCIS, Etude sur le potentiel du stockage d'énergie, October 2013.

[ADE 14] ADEME, Fiches des démonstrateurs Smart Grids du programme d'investissements d'avenir, June 2014.

[ALL 12] ALLIANCE G3, G3-PLC powerline communication standard for today's smart grid, October 2012. Available at http://www.g3-plc.com/

[ALL 14] ALLIANCE G3, Présentation de l'alliance CPL G3, 29 August 2014.

[CAP 12] CAPENERGIES, SCS, La chaîne de valeur du marché des smart grids, ITEMS International, June 2012.

[CEE 14] COUNCIL OF EUROPEAN ENERGY REGULATORS (CEER), "Status review on European Regulatory approaches enabling smart grids solutions 'Smart Regulation'", February 2014.

[CIG 14] CIGRE, Electric transmission and distribution power, Special issue on Russian Networks, August 2014.

[CRE 13] COMMISSION DE RÉGULATION DE L'ENERGIE (CRE), Développement des réseaux électriques intelligents en basse tension, 4 November 2013.

[CRE 14a] COMMISSION DE RÉGULATION DE L'ENERGIE, (CRE), "l'Expérimentation Linky en France", http://www.smartgrids-cre.fr/index.php?p=compteurs-linky, 2014.

[CRE 14b] COMMISSION DE RÉGULATION DE L'ENERGIE, (CRE), "Synthèse de la consultation publique de la CRE sur le développement des réseaux électriques intelligents en basse tension", http.cre.fr/documents/consultations-publiques/developpement-des-reseaux-electriques-intelligents-en-basse-tension, 12 June 2014.

[CRE 14c] COMMISSION DE RÉGULATION DE L'ENERGIE, (CRE), "Recommandations sur le développement des réseaux électriques intelligents en basse tension", http.cre.fr/documents/consultations-publiques/ developpement-des-reseaux-electriques-intelligents-en-basse-tension, 12 June 2014.

[DOE 12] US DEPARTMENT OF ENERGY (DOE), "2010 smart grid system report", February 2012.

[DOE 13] US DEPARTMENT OF ENERGY (DOE), "Smart grid investment grant program: progress report 2", October 2013.

[DOU 14a] DOULET A., Le développement des réseaux électriques de distribution en France, March 2014.

[DOU 14b] DOULET A., Quelques reflexions sur le développement des réseaux électriques de distribution à l'international, April 2014.

[DOU 14c] DOULET A., Smart Grids: une opportunité pour moderniser les réseaux de distribution, April 2014.

[EDP 14] EDP, "The InovGrid project", March 2014.

[EDS 14a] EDSO, EDSO presents views on system flexibility ahead of European Commission energy retail market communication, May 2014.

[EDS 14b] EDSO, Retail energy market: Supplementary response to the European Commission's public consultation, April 2014.

[EDS 14c] EDSO, Flexibility: The role of DSOS in tomorrow's electricity market, May 2014.

[EEA 13] EUROPEAN ENVIRONMENT AGENCY, Trends and projection in Europe 2013, October 2013.

[EIA 14] US ENERGY INFORMATION ADMINISTRATION (EIA), International energy outlook 2013, March 2014.

[ERD 13] ERDF, "Smart Grids, les réseaux électriques intelligents", July 2013.

[ERD 14a] ERDF, "Linky, le compteur nouvelle génération", http://www.erdf.fr/Linky/, 2014.

[ERD 14b] ERDF, En attendant les compteurs intelligents, 2014.

[ERD 14c] ERDF, "Spécifications du profil CPL Linky", http:// www. erdf.fr/medias/ Linky/ERDF-CPT-Linky-SPEC-PROFIL-CPL_ Specifications_du_profil_CPL_Linky.pdf, 2014.

[ENE 14] ENEXIS, International development of G3-PLC, March 2014.

[EUC 10] EU COMMISSION TASK FORCE FOR SMART GRIDS – EXPERT GROUP 1, Functionalities of smart grids and smart meters, December 2010.

[EUC 13] EU COMMISSION TASK FORCE FOR SMART GRIDS – EXPERT GROUP 3, Options on handling smart grids data, January 2013.

[EUC 14] EU COMMISSION – DG ENER, The role of DSOs in a smart grid environment, April 2014.

[EUR 13a] EURELECTRIC, Power distribution in Europe – facts and figures, November 2013.

[EUR 13b] EURELECTRIC, Active distribution system management: a key tool for the smooth integration of distributed generation, February 2013.

[EUR 14] EURELECTRIC, Flexibility and aggregation: requirement for their interaction in the market, January 2014.

[EWI 12] EWI, Flexibility options in European electricity markets in high RES-E scenarios, October 2012.

[GAL 11] GALLI S., SCAGLIONE A., WANG Z., "For the grid and through the grid: the role of power line communication in the smart grid", *Proceedings of the IEEE*, vol. 99, no. 6, pp. 998–1027, June 2011.

[GSG 12] GLOBAL SMART GRID FEDERATION (GSGF), "Report 2012", 2012.

[HAD 10] HADJSAID N. (ed.), La distribution d'énergie électrique en présence de production décentralisée, Hermès Science – Lavoisier, March 2010.

[HAD 11] HADJSAID N., SABONNADIÈRE J.-C., *Electrical Distribution Networks*, ISTE, London and John Wiley & Sons, New York, May 2011.

[HAD 12] HADJSAID N., SABONNADIÈRE J.-C., *Smart Grids*, ISTE, London and John Wiley & Sons, New York, April 2012.

[HAD 13] HADJSAID N., SABONNADIÈRE J.-C., *Power Systems and Restructuring*, ISTE, London and John Wiley & Sons, New York, February 2013.

[IEA 13] INTERNATIONAL ENERGY AGENCY (IEA), "Global EV Outlook – understanding the electric vehicle landscape to 2020", April 2013.

[IEA 14a] INTERNATIONAL ENERGY AGENCY (IEA), "World energy Outlook 2013", March 2014.

[IEA 14b] INTERNATIONAL ENERGY AGENCY (IEA), "Energy policy Highlights 2013", March 2014.

[IEA 14c] INTERNATIONAL ENERGY AGENCY (IEA), "Energy Technology Initiatives 2013", March 2014.

[ISG 13] ISGAN, AMI case book spotlight on advanced metering infrastructure, April 2013.

[ISG 14] ISGAN, Spotlight on demand side management, February 2014.

[JRC 14] JRC, "Smart Grid Projects Outlook 2014", June 2014.

[KPM 12] KPMG/INSTITUT CHOISEUL, Compétitivité énergétique des Etats, Baromètre 2012, 26 November 2012.

[LEW 12] LEWINER C., DAVID P., European energy challenges, 25 April 2012.

[MAL 14] MALLET P., GRANSTRÖM P.-O., HALLBERG P., *et al.*, "European perspectives on the future of electric distribution", *IEEE Power & Energy magazine*, vol. 12, pp. 51–64, 2014.

[MCK 13] MCKINSEY GLOBAL INSTITUTE, Disruptive technologies: advances that will transform life, business, and the global economy, May 2013.

[OBS 13] OBSERV'ER, Etat des énergies renouvelables en Europe, 2013.

[PER 84] PERSOZ H., LEMOINE J.-C., SAPET P., *et al.*, *La planification des réseaux électriques*, Eyrolles, 1984.

[PER 12] PERCEBOIS J., MANDIL C., Report energies 2050, 2012.

[ROC 02] ROCHET J.C., TIROL J., "Platform competition in two-sided markets", *Journal of European Economic Association*, vol. 1, no. 4, pp. 990–1029, 2002.

[RTE 14] RTE, "Bilan électrique 2013", February 2014.

[SAL 13] SALON C., HUET O., BLANC F., How DSO facilitate system optimisation in new market mechanisms for distributed resources, Internal Report, June 2013.

[UIT 12] UIT-T, Emetteurs-récepteurs de courants porteurs en ligne avec multiplexage par répartition orthogonale de la fréquence (OFDM) à bande étroite – Spécification de la densité spectrale de puissance, November 2012.

[WEC 12] WORLD ENERGY COUNCIL: WORLD ENERGY PERSPECTIVE, Smart grids: Best practice fundamentals for a modern energy system, 2012.

[WSG 13] WORLD SMART GRID FORUM 2013, Results and recommendations, September 2013.

[ZPR 11] ZPRYME RESEARCH & CONSULTING, China: rise of the smart grid, January 2011.

[ZPR 12] ZPRYME RESEARCH & CONSULTING, China: state grid corporation of China profile, March 2012.

Author articles and interviews

[BOI 11a] BOILLOT M., "La révolution annoncée du compteur communicant", *BIP Enerpresse*, 12 April 2011.

[BOI 11b] BOILLOT M., "ERDF: signature de contrat avec la Russie", *Le Figaro*, 18 March 2011.

[BOI 11c] BOILLOT M., "La grande évolution des réseaux d'électricité", *Revue de l'Electricité et de l'Electronique*, no. 6, 2011.

[BOI 12a] BOILLOT M., "Linky, un élément déterminant du système électrique", *Enerpresse*, 24 January 2012.

[BOI 12b] BOILLOT M., "ERDF doit faire face à la montée en puissance des énergies éolienne et solaire", *Smart Planet.fr*, 24 January 2012.

[BOI 12c] BOILLOT M., "Une Chaire d'excellence sur les Smart Grids", *Le Dauphine*, 6 May 2012.

[BOI 12d] BOILLOT M., "Smart, un peu, beaucoup, passionnement!", *Energie plus*, October 2012.

[BOI 12e] BOILLOT M., "ERDF: un fil conducteur, l'innovation", *Le journal des Grandes Ecoles*, April 2012.

[BOI 12f] BOILLOT M., Le "Smart Grid, réseau de demain", *La Jaune et la Rouge*, April 2012.

[BOI 13a] BOILLOT M., "La révolution smart grids: quels développements et perspectives au-delà des réseaux d'énergie?", *Forum syntec ingenierie*,7 November 2013.

[BOI 13b] BOILLOT M., "Réseaux intelligents, la révolution opportune", *Le Moniteur*, November 2013.

[BOI 13c] BOILLOT M., "Electricité: de la souplesse sur le réseau", *Science et Avenir*, October 2013.

[BOI 13d] BOILLOT M., "Comment optimiser le réseau", *Environnement et Energie*, June 2013.

[BOI 13e] BOILLOT M., "Houat et Hoëdic ont expérimenté la limitation télécommandée de puissance", *Enerpresse*, April 2013.

[BOI 13f] BOILLOT M., "SoGrid: la révolution du réseau pensant", *Le Point*, April 2013.

[BOI 13g] BOILLOT M., "ERDF: plus visible et plus exposé", *Enerpresse*, March 2013.

[BOI 13h] BOILLOT M., "En Vendée, le réseau électrique va devenir intelligent", *La Croix*, July 2013.

[BOI 13i] BOILLOT M., "Les missions tout-terrain d'ERDF au Liban", *Les Echos*, 6 March 2013.

[BOI 13j] BOILLOT M., "Réseaux électriques intelligents", *Le Monde*, 28 November 2013.

[BOI 13k] BOILLOT M., "Quand la réforme du réseau de l'électricité se met en route", *L'Orient Le Jour*, 23 January 2013.

[BOI 13l] BOILLOT M., "La gestion déléguée des réseaux, spécialité française", *CCE International*, January 2013.

Index

A, B, C

D, E

F, G, H, I

L, M, N

O, P

Q, R

S

T, U, V, W

Other titles from

ISTE

in

Electrical Engineering

2013

GALLOT-LAVALLÉE Olivier
Dielectric Materials and Electrostatic

GLAIZE Christian, GENIES Sylvie
Lithium Batteries and Other Electrochemical Storage

PERA Marie-Cécile, HISSEL Daniel, GUALOUS Hamid, TURPIN Christophe
Electrochemical Components

2012

BECKERS Benoit
Solar Energy at Urban Scale

FOULADGAR Javad
Electrothermics

GLAIZE Christian, GENIES Sylvie
Lead and Nickel Electrochemical Batteries

ROBOAM Xavier
Systemic Design Methodologies for Electrical Energy Systems

ROBOAM Xavier
Integrated Design by Optimization of Electrical Energy Systems

ROBYNS Benoît *et al.*
Electricity Production from Renewable Energies

SABONNADIERE Jean-Claude, HADJSAID Nourédine
Smart Grids

2011

GAO Fei, BLUNIER Benjamin, MIRAOUI Abdellatif
Proton Exchange Membrane Fuel Cells Modeling

SABONNADIERE Jean-Claude, HADJSAID Nourédine
Electrical Distribution Networks

LOUIS Jean-Paul
Control of Non-conventional Synchronous Motors

LOUIS Jean-Paul
Control of Synchronous Motors

MONMASSON Eric
Power Electronic Converters: PWM Strategies and Current Control

MULTON Bernard
Marine Renewable Energy Handbook

RAZIK Hubert
Handbook of Asynchronous Machine with Variable Speed

TRIGEASSOU Jean-Claude
Electrical Machines Diagnosis

ZAIM Mohammed El-Hadi, REZZOUG Abderrezak
Non-conventional Electrical Machines

2010

BERETTA Joseph
Automotive Electricity / Electric Drive Systems

BRUNET Yves
Energy Storage

DE FORNEL Bernard, LOUIS Jean-Paul
Electrical Actuators / Identification and Observation

JUFER Marcel
Electric Drive / Design Methodology

LE DOEUFF René, ZAÏM Mohammed El-Hadi
Rotating Electrical Machines

MARTINEZ-VEGA Juan
Dielectric Materials for Electrical Engineering

2009

HADJSAÏD Nouredine, SABONNADIÈRE Jean-Claude
Power Systems and Restructuring

HUSSON René
Control Methods for Electrical Machines

MOTTIER Patrick
LED for Lighting Applications

SABONNADIERE Jean-Claude
Low Emission Power Generation Technologies and Energy Management

SABONNADIERE Jean-Claude
Renewable Energy Technologies

2008

CRAPPE Michel
Electric Power Systems

MEUNIER Gérard
The Finite Element Method for Electromagnetic Modeling

2006

PRÉVÉ Christophe
Protection of Electrical Networks

Lightning Source UK Ltd.
Milton Keynes UK
UKOW07n2124120215

246171UK00009B/46/P

9 781848 217379

8 139.95